Manual of Water Supply Practices

M6

Water Meters: Selection, Installation, Testing, and Maintenance

Sixth Edition

American Water Works Association

Manual of Water Supply Practices—M6, Sixth Edition

Water Meters: Selection, Installation, Testing, and Maintenance

Disclaimer

The authors, contributors, editors, and publisher do not assume responsibility for the validity of the content or any consequences of their use. In no event will AWWA be liable for direct, indirect, special, incidental, or consequential damages arising out of the use of information presented in this book. In particular, AWWA will not be responsible for any costs, including, but not limited to, those incurred as a result of lost revenue. In no event shall AWWA's liability exceed the amount paid for the purchase of this book.

Manager – Manuals: Molly Beach
Manager – Publishing Operations: Gillian Wink
Specialist – Copyright and Permissions: Peggy Tyler
Director of Publishing: John Fedor
Technical Editor: Anne Lanctôt
Technical Illustrations: Michael Labruyere
Cover Image: Jarn Aui/Shutterstock.com
Production: Innodata

Library of Congress Cataloging-in-Publication Data

CIP data pending from the Library of Congress.

ISBN 978-1-64717-298-5 ISBN, electronic 978-1-61300-818-8

http://doi.org/10.12999/AWWA.M6ed6

American Water Works Association

American Water Works Association
6666 West Quincy Avenue
Denver, CO 80235-3098
awwa.org

Contents

Figures

Tables

This page intentionally blank.

Foreword

This manual is a guide to selecting, installing, testing, and maintaining water meters. The manual discusses recommended practices; it is not an AWWA standard calling for compliance with certain specifications. It is intended for use by new and existing utilities of all sizes, either as a textbook for those not fully familiar with meters or as a reference manual. Water utilities may use this manual as a tool to obtain the best revenue for their investment in meters and maintenance facilities. Design engineers and consultants may use this manual in laying out new meter installations or repair shops.

Most readers will first use this manual for assistance on an individual problem; however, for adequate knowledge of the field of water meter usage, the entire manual should be studied. This manual references applicable AWWA standards for the various types of meters. These standards may be purchased by calling AWWA Customer Service (800.926.7337) or through the Bookstore section of the AWWA website, awwa.org.

The AWWA manual on water meters was first published in a series of articles in the June, July, August, and September 1959 issues of *Journal AWWA* and was compiled as a separate reprint in 1962. A second edition of the manual was published in 1973, as a complete revision, updating and substantially enlarging the 1962 serial edition. The third and fourth editions, published in 1986 and 1999, respectively, reflected the latest revisions of the AWWA meter standards at the time and included substantial updates to the text and photographs.

The fifth edition, published in 2012, provided another update from the AWWA meter standards, including new AWWA standards on singlejet meters and fluidic-oscillator meters.

This sixth edition contains current information from the latest AWWA meter standards, including those for residential fire sprinkler meters (ANSI/AWWA C714) and electromagnetic and ultrasonic meters (ANSI/AWWA C715). Metering technology continues to grow rapidly, and the future holds even greater promise.

This page intentionally blank.

Acknowledgments

The AWWA Subcommittee on Revision of Manual M6, which developed this manual, had the following personnel at the time:

Michael L. Mastic, *Chair*
Peter Roberts, *Vice-Chair*

R.A. Barillas, Badger Meter, Milwaukee, Wis.
M. Beach (*AWWA staff advisor*), AWWA, Denver, Colo.
E. Cohee, MARS Company, Urbana, Ind.
M.D. Cole, Meter Technology Werks, Tampa, Fla.
J. Conover, City of Sacramento, Sacramento, Calif.
G.H. De Jarlais, Badger Meter, Milwaukee, Wis.
D.J. Devane, Zenner USA, Leeds, Ala.
A. Dudley, Kamstrup Water Metering LLC, Roswell, Ga.
D. Faber, Faber & Associates, Terra Alta, W.Va.
A. Frank, Neptune Technology Group, Inc., Tallassee, Ala.
R. Gallon, Zenner USA, Banning, Calif.
J.G. Gunn, Johnson Controls, Inc., Victor, N.Y.
C.C. Hannah, Johnson Controls, Inc., Lubbock, Tex.
A. Hendey, Sr., Hendey Meter, Beaumont, Calif.
D.M. Hill, American Water, Voorhees, N.J.
D.E. Hood, M.E. Simpson Company, Inc., Valparaiso, Ind.
A.M Horbovetz, M.E. Simpson Company, Inc., Valparaiso, Ind.
W.J. Jernigan, Cavanaugh & Associates, Winston Salem, N.C.
M.C. Johnson, Utah State University, Logan, Utah
M.J. Keilty, Endress+Hauser Flowtec, Estes Park, Colo.
T.A. Kelly, Martinsburg, W.Va.
R.N. Koch, Master Meter, Inc., Pittsburgh, Pa.
M.B. Kristensen, Kamstrup, Skanderborg, Denmark
A. Land, Dallas Water Utilities, Dallas, Tex.
L. Lewison, Cavanaugh & Associates, Mills River, N.C.
M.L. Mastic, MARS Company, Ocala, Fla.
P.W. Mayer, Water Demand Management, Boulder, Colo.
J.A. McCraven, Mueller Systems, Cleveland, N.C.
K. Molli, Menomonee Falls, Wis.
B. Murphy, Kamstrup Water Metering LLC, Roswell, Ga.
J.A. Olson, Badger Meter, Milwaukee, Wis.
D. Padron, AV Water Technologies, Houston, Tex.
J.E. Pintok, RG3 Meter, Lake Mary, Fla.
R.A. Richter, National Institute of Standards and Technology, Gaithersburg, Md.
P. Roberts, Ford Meter Box, Wabash, Ind.

J.R. Scarborough, Neptune Technology Group, Inc., Tallassee, Ala.
S.H. Seehoffer, Master Meter, Inc., Mansfield, Tex.
M. Shamley, Metron-Farnier LLC, Boulder, Colo.
C. Sharkey, Valor Water Analytics, Inc., Davis, Calif.
R. Sloan, Birmingham Water Works Board, Birmingham, Ala.
J. Solanes, Meter Technology Werks, Tampa, Fla.
S.M. Swanson, Sensus, Uniontown, Pa.
J. Szewczyk, Sensus | Xylem, Uniontown, Pa.

The AWWA Standards Committee on Water Meters, which reviewed and approved this manual, had the following personnel at the time of approval:

Craig C. Hannah, *Chair*
Michael L. Mastic, *Vice-Chair*
Monte R. Merkel, *Secretary*

General Interest Members

A. Cecchetti (*liaison, nonvoting*), Standards Council Liaison, San Francisco Bay Regional Water Quality Control Board, Oakland, Calif.
J.E. Cray, Wright-Pierce, Andover, Mass.
D. Faber, Faber & Associates, Terra Alta, W.Va.
R.C. Graff, Poway, Calif.
J.G. Gunn (*alternate*), Johnson Controls, Inc., Victor, N.Y.
C.C. Hannah, Johnson Controls, Inc., Lubbock, Tex.
D.E. Hood, M.E. Simpson Company, Inc., Valparaiso, Ind.
A.M. Horbovetz (*alternate*), M.E. Simpson Company, Inc., Valparaiso, Ind.
M.C. Johnson, Utah Water Research Laboratory, Utah State University, Logan, Utah
T.A. Kelly, Jr., Martinsburg, W.Va.
F.S. Kurtz (*liaison, nonvoting*), Standards Engineer Liaison, AWWA, Denver, Colo.
M.L. Mastic, MARS Company, Ocala, Fla.
M.R. Merkel, SL-serco, Inc., Evansville, Ind.
K.C. Molli, Menomonee Falls, Wis.
R.A. Richter, National Institute of Standards and Technology, Gaithersburg, Md.

Producer Members

R. Barillas (*alternate*), Badger Meter, Milwaukee, Wis.
T.D. Bianchi, Neptune Technology Group, Inc., Tallassee, Ala.
C. Brunson (*alternate*), Neptune Technology Group, Inc., Tallassee, Ala.
M.D. Cole, Meter Technology Werks, Tampa, Fla.
D.J. Devane, Zenner USA, Leeds, Ala.
M. Drake (*alternate*), Metron-Farnier LLC, Boulder, Colo.
R. Gallon (*alternate*), Zenner USA, Banning, Calif.
L. Gregory (*alternate*), RG3 Meter Company, Longview, Tex.

A. Hendey, Sr., Hendey Meter, Beaumont, Calif.
B. Kaffenberger, Itron, Liberty Lake, Wash.
R.N. Koch, Master Meter, Inc., Pittsburgh, Pa.
J.A. McCraven, Mueller Systems, Cleveland, N.C.
J. Olson, Badger Meter, Milwaukee, Wis.
J. Pintok, RG3 Meter Company, Lake Mary, Fla.
M. Shamley, Metron-Farnier LLC, Boulder, Colo.
J.R. Solanes (*alternate*), Meter Technology Werks, Tampa, Fla.
S.M. Swanson, Sensus, Uniontown, Pa.
J. Szewczyk (*alternate*), Sensus | Xylem, Uniontown, Pa.
W.J. Vetter (*alternate*), Master Meter, Inc., Mansfield, Tex.

User Members

M.J. Aragon, Denver Water, Denver, Colo.
M.C. Bowen, City of Columbus, Division of Water, Columbus, Ohio
J. Conover, Jr., City of Sacramento, Sacramento, Calif.
G.M. Gehringer (*alternate*), New Jersey American Water, Delran, N.J.
A.J. Haas, Arizona Water Company, Phoenix, Ariz.
P.A. Hayes, Mammoth Community Water District, Mammoth Lakes, Calif.
D.M. Hill, American Water, Voorhees, N.J.
N.D. Kaufman, Truckee Donner Public Utility District, Truckee, Calif.
P. Kostelec, Dallas Water Utilities, Dallas, Tex.
A. Land, Dallas Water Utilities, Dallas, Tex.
R. Molhoek, Desert Water Agency, Palm Springs, Calif.
G.J. Oakley, Philadelphia Water Department, Philadelphia, Pa.
C. Patterson, Washington Suburban Sanitary Commission, Laurel, Md.
B. Schade, WaterOne, Lenexa, Kans.
R. Sloan, Birmingham Water Works Board, Birmingham, Ala.
D. Strub, Austin Water Utility, Austin, Tex.

The AWWA Standards Committee on Water Meters acknowledges the contributions and leadership of George H. De Jarlais, who was the Manual M6 subcommittee chair during the development stage for new content for this manual. The committee also acknowledges Andrew T. Dudley, who provided major contributions to this sixth edition of the manual before his untimely passing in 2024. Mr. Dudley was an emerging leader in AWWA meter standards work, in addition to being a mentor and friend to all.

This page intentionally blank.

Chapter 1

History of Water Measurement and Development of Water Meters

INTRODUCTION

Accurate water measurement is the means by which water utilities produce revenue to cover expenses, charge each customer equitably, manage demand, prevent waste of water, and minimize the load on wastewater facilities. This concept is universally accepted today, but it took thousands of years for the science of water measurement to reach its present state.

WATER SUPPLY IN ANCIENT TIMES

Until about 313 B.C.E., the citizens of Rome were content to draw their water supply from the Tiber River, wells, and springs. As in our own times, such sources became entirely inadequate to meet the needs of a rapidly growing city, and the first of the great aqueducts, the Appian, was constructed. This was followed 40 years later by the Anio Vetus and during the next three centuries by seven other aqueducts.

All of these aqueducts are described in meticulous detail by Water Commissioner Sextus Julius Frontinus (approximately C.E. 35–104) in his two books of *De Aquis Urbis Romae* (translated into English by Clemens Herschel under the title *The Water Supply of the City of Rome*). The contributions made by Frontinus are significant to the history of water engineering and beyond. The office of water commissioner was held, from ancient times, by the most distinguished citizens. Frontinus set an example for all future water commissioners to follow in his honest devotion to duty, thoroughness of accomplishment, and sterling character. Those who read his voluminous report today can still profit from it and are likely astonished by the progress made during those ancient times in the management of water systems through the development of metal pipe, pipe fittings, plug cocks, inverted siphons, ornate fountains and baths, and possibly in methods of measurement and filtration. This is the earliest known origin of public supply water measurement and management.

MECHANICAL WATER MEASUREMENT AND METERING

As cities and urban water systems expanded across the world in the 1700s and 1800s, it became increasingly important to accurately measure water volumes for the purpose of charging for use and to reduce waste; thus, water meters were invented. Most metering advances over the past 250 years were driven by the desire to eliminate waste and better account for scarce resources. Even today, modern water meters are promoted for their increased accuracy and robust performance versus previous models. In addition, utilities have continually sought improved metering solutions to ensure full revenue collection and accountability.

Pitot, Darcy, and Company

In 1730 on the banks of the Seine River in Paris while conducting research into the dynamics of incompressible fluids, the French engineer Henri Pitot (1695–1771) used a vertical glass tube with the short, 90° tip at the lower end pointing upstream to determine a relationship between the rise of water in the tube and the velocity of flow. In 1732, he published a paper in *l'Academie Royale des Sciences* describing his experiments and his tube, which he had combined with a second, static-head vertical tube.

Pitot's discovery that the height to which the water rose in his tube was proportional to the square of the stream velocity came in the same year that Johann Bernoulli published the fundamental relationship of head to velocity squared of water flowing through pipes. These are the fundamental relationships of fluid mechanics.

In the 1840s and 1850s, French engineer and chief director Henry Darcy, another French engineer, modified the tube and thus improved the accuracy of the static-head readings and reduced the oscillations of the water columns. Darcy also modified the Prony equation for calculating head loss due to friction, which, after further modification by Julius Weisbach, would become the well-known Darcy–Weisbach equation still in use today.

In 1875, Hiram F. Mills of Boston became the first American engineer to use the Pitot tube, and John R. Freeman began using a Pitot tube to measure fire-stream jets for fire suppression. The invention of the pitometer by Edward S. Cole in 1896 made it possible for him to document the extent of water loss (waste) in water systems in the early 20th century.

Mechanical Meters: Piston, Woltmann, Turbine, Multijet

In 1790, Benjamin G. Hoffman, in Hamburg, Germany, published a booklet describing a meter, invented by Reinhard Woltmann (1757–1837), for measuring flowing air and water. This device may be the first practical water meter created.

In 1824, Thomas Kennedy of Argyleshire, UK, assisted by clockmaker John Cameron, made a single-piston, double-acting water meter. Kennedy meters were known in the United Kingdom and United States (Crainic 2012).

In the United States, while patents for liquid meters were issued earlier, it is believed that the first American water meter was proposed in 1850, by William Sewell of Williamsburg, N.Y., and the first meter actually produced is claimed to be made by Henry R. Worthington of New York. These early meters were of reciprocating piston design, fashioned after the familiar D-slide valve used in steam engines (Crainic 2012).

In England, the first turbine water meter was designed in 1851 by William Siemens (1823–1883). Siemens obtained permission to manufacture his meter in Germany, and the firm of Siemens-Halske began production about 1865. Although such meters appeared to give fairly accurate results, corrosion caused considerable trouble because the meter casings were made of cast iron and the internal working parts of bronze.

Woltmann turbine meters were described in a German handbook, *Die Hydraulik Und Die Hydraulischen Motoren*, by Georg Meissner, published in Jena, Germany, in 1878. In this era, it was not generally accepted that a turbine meter could be used in a closed pipe. The 1885 book *Water Meters* by Ross E. Browne described the English Siemens ½-in. (13-mm) meter, which somewhat resembled a centrifugal pump reversed, using a drum reaction wheel controlled by regulating vanes. The German Siemens device was an early form of the rotary-vane meter.

In the United States, spurred by the rapid adoption of the toilet, an increased demand for water, and the desire for reliable urban supplies, more than 400 patents were issued on liquid meters between 1870 and 1910. Many of these patents describe devices that were hopelessly impractical and complicated, but one of these patents issued to J.P. Lindsay in 1871 described an oscillating disc meter, which today would be called a reciprocating meter with a single piston. In 1884, the first patent for a practical oscillating-piston meter was issued to Lewis H. Nash and described as an adaptation of an existing steam engine, but the piston shape and movement are similar to the oscillating-piston meters of today. The Nash device was a simple, workable measuring unit (Crainic 2012).

The origin date of the first nutating-piston or disc meter is not clear, but it is known that these devices evolved from hydraulic engines and engine valves built from the 1820s to the 1940s. Limited patents on disc-type meters were issued in 1887 and 1888, and patent issued in 1892 to James A. Tilden covered features of a flat-disc meter. Many meter designs from this period were adopted and developed into companies that are ancestors (or brands) that still have operating units producing liquid water meters.

The Venturi Meter

A major contribution to the measurement of flowing water under pressure was Clemens Herschel's invention of the Venturi tube in 1886. A hydraulic engineer with the Holyoke (Mass.) Water Power Company, he was disturbed by the fact that huge quantities of water used by industrial companies, including 25 paper mills, went unmeasured. The tube was based on the principle, discovered a century earlier by the Italian engineer Giovanni Venturi (1746–1822), that the flow of liquid through a converging pipe results in a gain of velocity and a lowering of pressure, whereas the reverse is true in a diverging pipe. The reduction in pressure that occurs at the constriction is actually the result of conversion of pressure head to velocity head. Downstream from the constriction, the velocity head is reconverted to pressure head, so the ultimate result is very little reduction in effective line pressure at some distance downstream. The difference between the full-pressure upstream taps and the reduced-pressure taps at the constriction of the throat or nozzle is a function of the flow rate and may be used to calculate the flow rate.

The Venturi meter has no moving parts in the stream and is suitable for nearly any size pipeline. One disadvantage of this meter is its length, which makes it bulky and heavy. Recently, however, shorter forms of the Venturi meter, known as flow tubes and flow nozzles, have been developed. These designs have a more abrupt constriction and enlargement. Also, the pressure taps are located differently, which greatly reduces overall length and weight.

One form of mechanical readout is a pressure-actuated pen arm moving across a graduated scale to indicate flow rate or across a moving chart to record the flow rate. More modern versions of the chart recorder are electronic. An integrating attachment may be added that multiplies the pressure differential by time and then multiplies that result by a constant or coefficient for the particular size of Venturi to register total flow within commercial tolerances.

The Orifice Meter

The orifice meter is considered to be one of the oldest devices for measuring or regulating the flow of fluids, dating back to Julius Caesar's Rome. However, development of the thin-plate orifice, the current standard type, did not begin until the early part of the 20th century.

The determination of the relationship between flow rates and differential pressures has been reduced to formulas by which an orifice coefficient can be computed quickly for any size of orifice in any size of pipe.

Modern differential pressure measurement instruments have an electronic output for recording instruments and incorporate electronic microprocessors for calibrated integration of the differential input signal from the meter. The electrical output signal is linear versus flow rate for flow indicators and data loggers.

The orifice sensing unit consists of a round plate with a hole of predetermined size bored in it, usually concentrically; the plate is inserted across a straight run of pipe. Pressure taps are provided either at specified distances upstream and downstream from the orifice plate or within orifice flanges by which the orifice plate is inserted in the line. The pressure differential created by the orifice is converted into linear motion of a pen arm, or integrator cam, by a manometer, which may be either a mercury or a sealed-bellows type. By calculating the orifice coefficient and drilling the orifice bore to the diameter required, the chart indicator, scale indicator, or integrator can be read in gallons or other units. This procedure is essentially the same whether a Venturi, an orifice, or some other pressure differential meter is used.

Orifice meters are used where space for the sensing unit is too limited for a Venturi meter and where slightly greater pressure loss can be permitted. The range of flow for maximum accuracy is about 3½ to 1. The sensing unit is relatively inexpensive. Orifice meters are widely used in industrial plants for water, steam, and gas and by gas utilities.

One water-utility application is to measure the flow of water from pumped wells being tested for flow rates. This operation is feasible provided that

- the pressure differential between the upstream and downstream taps is not more than the limit of the bellows manometer (maximum limit approximately 14 psi [100 kPa]),
- the orifice plate and flanges fit the pipe correctly, and
- the discharge leading to the atmosphere can be kept under slight back pressure.

Displacement Meters

The most popular types of displacement meters originated as modifications of pumps that were run backwards. Instead of using the mechanism as a prime mover, the water under pressure moved the piston or disc. The number of strokes or the measured volume per cycle was converted into gallons, cubic feet, or other unit of measurement by appropriate gearing.

In the 1850s, the George Kent Scientific & Mechanical Group of Luton, UK, manufactured a duplex-piston meter of nearly identical design. Samples of this meter exist in several European museums and meter displays in the Chicago and Baltimore meter shops.

These early meters were of reciprocating piston design, patterned after the familiar D-slide valve used in steam engines. Worthington's original meter consisted of two cylinders and plungers with inlet and outlet ports arranged in such a way that as water in one cylinder was discharged by the piston, the other cylinder was filling. Water flow through these meters was subject to pulsations caused by the action of the pistons, and high friction losses were encountered. An analysis of the ⅝-in. (15-mm) Worthington meter, made in 1885 by Ross E. Browne, describes it as having a capacity of 7.5 gal per min (gpm) (1.7 m^3/h), weighing 59 lb (27 kg), and costing $17.00.

Close accuracy was impossible because the meter counted the number of piston strokes, whose length varied with the flow rate. Variations in velocity of the strokes caused variable compression of the rubber cushions at each end of the cylinders.

In 1876, a patent was issued for a meter with two rotary pistons. A design for only one piston followed around 1880. The piston rotated within a toothed, fixed cylinder called a crown. In 1885, this design was manufactured in France as the Nasch Crown meter and was also sold in the United States.

Browne also described a single-piston, double-acting meter, the Kennedy. In 1885, it weighed 162 lb (73 kg) in the ½-in. (13-mm) size and sold for $27.83. It measured the travel distance of the piston.

A nutating-piston liquid pump, using what today would be called a conical disc, was patented in United States in 1854 by Stephen D. Carpenter. A modification, also for use as a pump, was patented by Charles H. Hersey in 1868 (Crainic 2012).

The date of the first nutating-piston or disc meter is not clear. Limited patents on this type were issued in 1887 and 1888, and patent 486,992, issued in 1892 to J.A. Tilden, covered features of a flat-disc meter. In that same year, two companies that are still in the meter business today were organized, one to manufacture conical-disc meters and the other flat-disc meters. A third company was already making conical-disc meters.

All these manufacturers had problems with hard rubber discs breaking at fast flows. One solution—developed by John Thomson in his patents 476,102; 485,437 (1892); and 535,641 (1895)—was for a thrust roller to control the circumferential thrust of the disc in its movement and thus prevent jamming of the disc against the diaphragm at the slot. This method continues to be widely used.

Another approach was to mold a metal reinforcement into the hard rubber disc section of the meter. The diaphragm was then used as a thrust plate to prevent rotation. In all these designs, the disc section (whether flat or conical) was made as thin as possible to reduce weight; however, this necessitated a very close fit between the edge of the disc and the chamber wall to minimize slippage. In 1892, G.B. Bassett designed and began manufacturing ⅝-in. (15-mm) conical-disc meters with the disc section ¼-in. (6-mm) thick, instead of 3⁄32 in. (2 mm) or less. He found the added weight to be no apparent disadvantage, but the greater thickness gave a more perfect capillary seal against slippage for a given clearance. A reinforcing plate was later added.

The early use of meters was largely in the north, where freezing troubles were frequent. In cold weather, many homeowners drained water pipes at night. The conical-disc meter was favored at this time because it could be drained more readily than the flat-disc type. To prevent frost damage when pipes were not drained, two features were developed. John Thomson's patent 520,197 of 1894 covered a breakable frost bottom to be "blown out by the expansive action of freezing." His patent 771,337 of 1904 covered the use of bolts that would "break, strip or yield resiliently" under pressure greater than normal but yet within the strength of the casings.

Protection of the reducing gear train against wear by enclosing the gears in a grease-filled chamber came into widespread use following World War I. The grease-filled chamber was actually a very old device. The French Nasch meter of 1885 had this feature, as did the English Siemens. Other early patents implied it, without claiming the chamber as invention. The oil-enclosed gear train remained standard until the advent of magnetic-drive meters between 1957 and 1960.

New magnetic materials that retained high magnetic power indefinitely and were inexpensive were developed, which led to a major change in small-meter design. In the late 1950s, the magnetic drive was developed by several companies, most of which used small four-pole and later two-pole, usually round, ceramic magnets. The use of these magnets eliminated the stuffing box and gear train, transmitting the motion through the top of the bronze casing by magnetic flux. By 1970, almost all small meters built in the United States and many large ones were of this design. An example of a displacement type meter is shown in Figure 1-1. This meter has a permanently sealed register unit with no changeable calibrating gears. A cutaway of a displacement-type nutating-disc meter with magnetic drive is shown in Figure 1-2.

Present standards for displacement meters are ANSI/AWWA C700, Cold-Water Meters—Displacement Type, Metal Alloy Main Case, and ANSI/AWWA C710, Cold-Water Meters—Displacement Type, Plastic Main Case.

Fluidic-Oscillator Meters

In these meters, water enters the fluidic-oscillator through a nozzle that forms an accelerated jet. When the jet enters the flow chamber, it initially is drawn to one of the two diffuser walls. The jet travels along the wall and exits the flow chamber. A small portion

Source: Neptune Technology Group

Figure 1-1 A $^{5}/_{8}$-in. positive displacement meter

Figure 1-2 Displacement-type nutating-disc meter with magnetic drive

of the flow migrates through the feedback channel and impinges on the incoming jet. This causes the jet to flip to the other side of the chamber, where the process is repeated. This principle is known as the Coandă effect. The oscillation between the diffuser walls is continuous while flow is present, with the frequency of oscillation proportional to the volume of water moved through the chamber. The oscillation is sensed by electrodes placed next to each diffuser wall. An electrical current is induced in the jet from which the rate of oscillation is calculated and, over a period of time, a total flow is derived.

The standard for fluidic-oscillator meters is ANSI/AWWA C713, Cold-Water Meters—Fluidic-Oscillator Type.

Singlejet Meters

Singlejet water meters were initially developed in Europe in the early 1900s. Singlejet meters have been widely manufactured and used worldwide since that time but were not introduced into the North American market until the early 1990s.

A singlejet meter is a velocity-type meter that incorporates a single tangential jet. A tapered or full-size throat targets the water stream across a vertical shaft impeller. The impeller is offset from the water flowing through the tangential jet such that the water travels perpendicular to the rotational axis of the impeller. This is the characteristic that separates the singlejet meter from a typical turbine meter where water flows along the rotational axis rather than perpendicular to it. By operating perpendicular to the impeller, water flow through the singlejet meter exerts very little force on the bearings.

The standard for singlejet meters is ANSI/AWWA C712, Cold-Water Meters—Singlejet Type.

Multijet Meters

Multijet meters use multiple ports surrounding an internal chamber to create jets of water that move an impeller. The rotation speed of the impeller measures the velocity of water flow. Multijet meters were first designed and produced in Germany in 1867 beginning with Siemens-Halske. At least one United States meter manufacturer marketed a multijet meter in late 1920s and into the 1930s for export (Crainic 2012).

The standard for multijet meters is ANSI/AWWA C708, Cold-Water Meters—Multijet Type.

Turbine Meters

Turbine-type meters typically have blades on an internal rotor. The turbine blades housed in the chamber rotate as the water flows through it. A turbine water meter works by using the kinetic energy of flowing water to spin a rotor with angled blades where the rotational speed of the rotor is directly proportional to the flow velocity of the water, allowing the meter to calculate the volume of water passing through based on how fast the rotor spins. Although the turbine rotor has historically been directly coupled by gearing to a mechanical register, other means of rotor sensing have been used successfully. The total number of pulses gives a measure of the total flow. Figure 1-3 shows a 3-in. Class II turbine water meter. Figure 1-4 shows a Class II turbine meter cutaway.

The standard for turbine meters is ANSI/AWWA C701, Cold-Water Meters—Turbine Type for Customer Service.

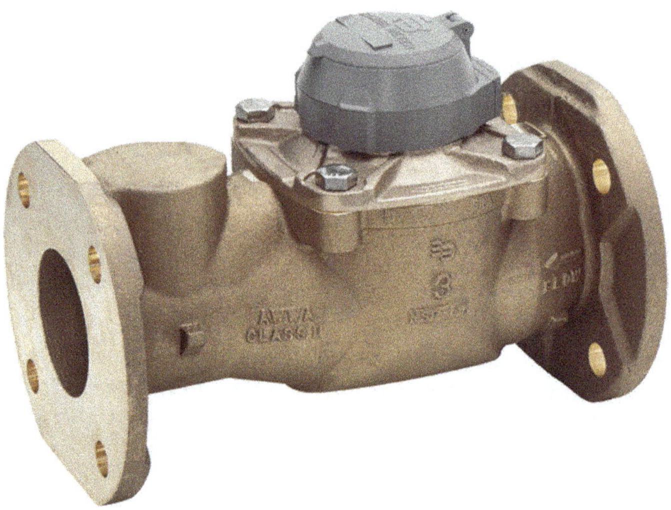

Source: Badger, Inc.

Figure 1-3 A 3-in. Class II turbine water meter

Source: Badger, Inc.

Figure 1-4 Class II turbine meter cutaway. Note the test port on top used for pressure testing, air bleeding, and qualitative low-flow tests

Compound Meters

The early history of meters included the development of various devices aimed at improving accuracy over a wide range of flows, including very low flows. This work resulted in the basic idea of a large meter on the mainline, a small meter on a bypass, and a valve to direct water to one meter or the other automatically. The registration was the total of the readings on the two meters on some designs and dual registers on other designs. The single-register design was covered in J.A. Tilden's initial patent 742,136 of 1903.

In 1909, F.N. Connet's patent 934,504 covered a single-register compound meter with a pair of ratchet-wheel drives arranged such that the register was driven by whichever drive moved faster. This patent also showed the mainline meter mounted horizontally. It was the direct forerunner of the latest high-flow single-register compound meters.

Much developmental effort has gone into the compounding valve to facilitate rapid opening and minimize loss of registration when changing from bypass to mainline registration. Modern swing-type and overcenter movement valves require a pressure differential (developed by flow through the bypass meter) to start to open but then move very easily to the full-open position. It is claimed that the single-register compounds suffer minimal changeover loss of accuracy as flow moves through the bypass and the mainline meters in series when the valve is open; therefore, whichever unit records the greater flow drives the register. Even though the single-register meter is easier to read, it must be inspected more frequently because trouble in one unit might not be readily apparent to the meter reader. The standard for compound meters is ANSI/AWWA C702, Cold-Water Meters—Compound Type. An example of a compound water meter, double register type, is shown in Figure 1-5.

Source: Neptune Technology Group

Figure 1-5 A compound meter with two registers

Fire-Service Meters

Water utilities have found it necessary to meter lines that are typically used for everyday needs and are occasionally required to supply water for firefighting purposes. Insurance underwriters have demanded that such lines have no restriction that could interfere with full flow of water in the event of fire, even if the meter is jammed or damaged.

To meet this requirement and still provide metering, a meter was developed that is a combination of a proportional mainline meter and disc or compound bypass meter, as described in Chapter 2. Several brands of meters have been approved for such use by UL Solutions and Factory Approvals. The first of these meters was approved in 1916.

The first known design was developed by J.A. Tilden, and the first meter was shipped by his company in 1908. His patent 971,510 was issued in 1910 and is similar to the construction now used. Another idea, patented in 1913 by J.R. Freeman (patent 1,057,721), used a Venturi-type, mainline meter. H. Chrisman's patent 1,118,921 of 1914 covered a displacement meter for measuring small quantities. It was positioned in the mainline but was swung out of the flow stream by a high flow rate so that the line was free of obstruction, but the meter was inoperative. In 1915, patent 1,138,640 by H.I. Dilts was the basis of the other approved fire-service meter, which used a disc-type bypass meter and a disc-type proportional meter for high flows.

For installations where emergency mainline flow is not to be metered, the detector check assembly is an option. The detector check consists of a weight-loaded valve in the mainline that remains closed under normal usage and a bypass around this valve that contains a water meter to measure only unauthorized use in a facility where utility access is normally limited. If there is enough daily use of water for restrooms, breakrooms, or kitchens, a correctly sized, separate service line and meter should be applied.

ANSI/AWWA C703, Cold-Water Meters—Fire-Service Type, provides specifications for fire-service meters. Many of the commercial-sized AWWA C715 meters have achieved insurance underwriter listing approval for use in combination fire-service and mainline applications. An example of a large fire-service meter configuration is shown in Figure 1-6.

Source: Neptune Technology Group

Figure 1-6 An example of a fire-service meter configuration

With the introduction of residential fire sprinkler systems, there are requirements for fire-service meters for smaller line sizes and with reduced performance requirements. Such meter applications are recognized in ANSI/AWWA C714, Cold-Water Meters for Residential Fire Sprinkler Systems in One- and Two-Family Dwellings and Manufactured Homes.

Proportional Meters

The proportional meter measures a relatively constant but small percentage of the water flowing through the line. A multiplying factor is built into the meter register to record the total quantity that has passed through the mainline. This task is accomplished by providing an expanded section of the line with a ring at the downstream end having an internal diameter equal to the nominal pipe diameter. The restriction ring creates a drop in line pressure just below it as compared with full-line pressure upstream. This differential in pressure forces a proportion of the water through the small displacement-type bypass meter.

When correctly calibrated, a proportional meter is accurate within 3 percent over a 10-to-1 or 12-to-1 total flow range, depending on the size of the bypass meter relative to the mainline. No AWWA standard covers only proportional meters because they are most commonly used as a fire-service meter.

Propeller Meters

The propeller meter was developed to permit self-contained measurement of water velocity in large lines up to 72 in. (1,800 mm), measure water containing impurities that might clog other types of meters, provide a mechanical meter that would operate with an absolute minimum loss of pressure, and permit saddle-mounting of the meter on a pipeline if required. These are technically classified as flowmeters used for control/monitoring purposes and are not normally used for billing purposes.

Accuracy provisions for propeller meters are contained in ANSI/AWWA C704, Propeller-Type Meters for Waterworks Applications.

Magnetic-Drive Meters

During the early use of water meters, there was a need to eliminate the accumulation of mud and fogging of the glass in register compartments of meters in pits or curb-box settings. The ventilation hole in the register compartment through which the stuffing-box leakage could escape also permitted moisture to condense beneath the glass and permitted mud to enter when meter pits were flooded by surface water. Reading of the meter was then impossible.

Three methods were tried to overcome this problem. During the 1930s and 1940s, some meters were built with the register glass capable of being lifted to read the dial beneath, but even more dirt could accumulate in the register compartment with this design. In 1917 and again in the late 1940s, a design was developed that placed the register in a sealed compartment, with a small stuffing box on the register drive shaft. The register unit was then filled with oil. This device helped, but leakage of the oil was difficult to prevent.

In 1912 and again in 1934, attempts were made to introduce meters with magnetically driven sealed registers. Both meters were discontinued because of the undependable magnets available for the drives, which were not permanent. In 1956, the first of the modern magnetically driven meters was introduced. It was an oscillating-piston meter with a permanently sealed register unit containing both register and reducing gear train. The magnetic drive eliminated the stuffing box and shaft previously used to transmit motion from the measuring piston to the register and was made possible by the newly developed permanent magnets. A magnet mounted in the drive pin of the oscillating piston traced a circular path around a well extending downward from the register unit. A free follower, another permanent magnet, rolled around on the inside of the well, pushing a crank that drove the register unit.

During the next few years, practically all US meter manufacturers released magnetically driven, sealed register meter designs. In most cases, the reducing gear trains were removed from the inside of the main casing. In these meters, the pin in the center of the disc or piston engages a driver block centrally mounted on a shaft in the driver magnet unit. On the upper end of the shaft, a ceramic-face magnet is mounted. The rotative motion of the driver magnet is transmitted to a driven magnet of similar design through the bronze main casing, where it drives the gearing and register.

This design has two styles. In one style, the register assembly is permanently factory-sealed, and no change gears are used. In the other, the register compartment can be opened in the meter shop for repair or for recalibration by the change gears. Both styles are considered sealed because, when assembled, the register is sealed against air and moisture, although change gear–equipped registers are obsolete today due to high maintenance cost. In the latter style, a capsule containing a desiccant is placed in the register compartment to absorb any moisture in the air at the time the meter is assembled. In some designs, shields of soft magnetic metals protect the magnetic couple against external interference.

The result has been meters that are more sensitive to low flow and have a longer service life. Friction load on the measuring unit has been reduced by eliminating the heavy intermediate gear train and the packing gland or stuffing box and by the extensive use of low-friction, lightweight parts in the register. The sealed register, with its high-impact-resistant lens, has reduced both fogging and breakage of the lens and corrosion of the working parts and has allowed field replacement of the register without interruption of service.

Electromagnetic and Ultrasonic Meters

Static meters include meters that do not have moving parts in the body of the meter. The newer static meters are designed as either electromagnetic meters or ultrasonic meters.

Electromagnetic meters use principles from Faraday's law of induction to calculate flow. An electromagnetic coil is used inside the meters to generate a magnetic field where an electromotive force can be measured and used to calculate the velocity profile and flow rate. The electromagnetic coil does not extend through the body of the meter, and therefore, there are no obstructions as water passes through electromagnetic meters.

Ultrasonic meters, another type of static meter suitable for specific measurement applications, use sound waves to pass through the body's measurement section of the meter to determine flow rate. Two or more paired points in the measurement section are used to measure travel time for sound waves to move with or against the fluid movement. Resistance from the water will affect the travel time of the sound wave, and this differential in transit time between the upstream path and the downstream path allows for a wide range of flow to be measured. Examples of ⅝-in. ultrasonic water meters are shown in Figures 1-7 and 1-8. Strainers are sometimes built into the bodies of electromagnetic and ultrasonic meters, as well as shut-off valves in smaller meters. ANSI/AWWA C715, Cold-Water Meters—Electromagnetic and Ultrasonic Type, for Revenue Applications, provides specifications for these types of static meters.

Source: Badger, Inc.

Figure 1-7 A 5/8-in. ultrasonic water meter

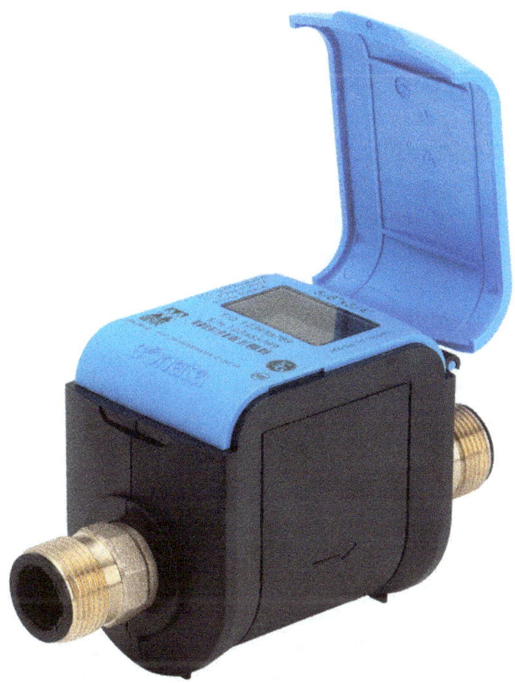

Source: Master Meter

Figure 1-8 A 5/8-in. ultrasonic water meter

ACCEPTANCE OF METERS BY WATER UTILITIES

The practice of charging for water by metering was questioned by water utilities for several years. One of the chief objections was that occupants of metered homes would attempt to keep water charges at a minimum by not using enough water for sanitary purposes. It was even reported that some users placed a tub under a faucet and allowed water to drip into the tub at a rate too slow for the meter to register. Water was dipped from the tub with a pail as needed.

This problem was solved by establishing a monthly minimum charge for which enough water was furnished to provide for reasonable sanitary needs. No carryover of unused allowance to the next month was permitted. This method opened the way to universal metering once improved meters were available, and it was determined that the total cost of operating a metered system was less than the cost without meters because water usage was reduced.

STANDARDIZATION

Standardization of meters in the United States began in 1913. Fortunately, standardization issues were considered at this time because otherwise a chaotic situation similar to one existing abroad might have resulted. An AWWA committee, appointed in 1913, proposed the adoption of standards on overall lengths and connections for disc meters in 1915 and 1916. In 1917, the New England Water Works Association (NEWWA) also proposed a set of standards. No official action was taken, however, on either proposal. In 1916, the meter manufacturers organized a meter standards committee, on which most manufacturers were represented.

On March 9–10, 1920, the AWWA and NEWWA committees met jointly to review drafts of a proposed standard developed by the manufacturers' committee. The final draft of the "Standard Specifications for Cold-Water Meters, Disc Type" was approved by AWWA and NEWWA in 1921. A revision was adopted by AWWA in 1946 and another in 1964. An important feature of the 1946 revision was the adoption of exact thread specifications for spuds on ⅝-in. (15-mm) to 1-in. (25-mm) meters and for coupling nuts. These specifications differ slightly from general thread specifications. They were selected to permit full interchangeability of new meters and/or couplings with old meters or couplings having various off-standard threads.

The title of the 1946 edition was broadened to displacement meters to include the oscillating-piston meter, and the 1964 edition is designated as a standard rather than a specification. The revisions since 1971 include magnetic-drive meters and a ½-in. meter size. Standards for compound, fire-service, and current meters were first adopted in 1923; propeller meters were adopted in 1949; multijet meters were adopted in 1976; and plastic main-case meters were adopted in 1988. Standards for singlejet meters and fluidic-oscillator meters were adopted in 2002 and 2005, respectively. Standards are revised or reaffirmed at least every 5 years.

AWWA Manual M6 was first published in 1962 with significant updates in 1973, 1986, 1999, and 2012. In 1975, the first edition AWWA Manual M22, *Sizing Water Service Lines and Meters,* was published, codifying the basic meter and service line sizing approach still in use today. Subsequent editions of AWWA Manual M22 were published in 2004, 2014, and 2024.

Capacity and Pressure-Loss Ratings

Early meters had high pressure loss for a given capacity. This loss was partly due to their complicated design and partly deliberate to reduce the possibility of running the meter too fast because damage to parts at excess speed was likely. The problem persisted even after the present simple disc or oscillating-piston designs were developed. Discs could break at excess speed, so some meters were fitted with different chamber settings to compensate for high line pressure.

The first AWWA specifications, now known as standards, were adopted in June 1921 and addressed disc meters. The manufacturers who aided the standards committee had to develop a workable, uniform standard for pressure-loss limits and capacity ratings. At the suggestion of R.S. Bassett, the ratio of the capacities of the various sizes was fixed by setting a constant peripheral speed of a point on the edge of the disc in its up-and-down motion as the basis of measure of relative capacity. Because the ⅝-in. (15-mm) meter was already universally rated at 20 gpm (4.5 m^3/h), the rated capacities of the other sizes were set so that this speed was constant for all meter sizes. For the rated capacity of other types, the flow velocity of the liquid at the safe maximum operating capacity (SMOC) is a determining factor; for example, 20 gpm in ⅝-in. size residential meters roughly corresponds to a water velocity of 21 ft/s in the service line. Today, the SMOC is understood as the maximum intermittent flow rate through the meter. In the case of mechanical meters, it should only occur for short periods of time and at infrequent intervals. In the case of solid-state (electronic) meters with no moving parts, no limitation applies to the timeframe of flow rates at or below the SMOC.

The standards committee also agreed on a relatively uniform pressure loss for the different types within the 20-psi (140-kPa) pressure-loss limit adopted at that time. Although this limit was too high for average use, meters seldom run at more than 50 percent of rated capacity. As the pressure loss increases with the square of the increase in rate of flow, a meter with a pressure loss of 20 psi (140 kPa) at safe rated capacity had a loss of only 5 psi (35 kPa) at 50-percent rating, which was permissible. Later water meter design improvements greatly reduced pressure loss. Average pressure loss in new ⅝-in. (15-mm) meters today is less than 15 psi (104 kPa) at 20 gpm (4.5 m^3/h).

Size Designations

The size of a water meter has always been described in terms of the size of pipe for which the meter was originally intended. When meters started to become popular in the United States, the standard pipe for residential services was lead pipe with an internal diameter of ⅝-in. (15-mm) and wiped solder joints. Consequently, meters were built with ½-in. (15-mm) inside diameter (ID) inlets and outlets. To permit a meter to be removed for servicing, its ends were threaded with ¾-in. (20-mm) straight external threads, and couplings of the same ⅝-in. (15-mm) ID were used. These couplings had a ¾-in. (20-mm) threaded nut on one end to fit the meter and a smooth tinned end on the tailpiece suitable for making the wiped joint to the pipe.

With the advent of galvanized-iron pipe services around the 1890s (now typically galvanized steel), the standard residential service line used pipe with a nominal size designation of ½ in. This pipe still had a ⅝-in. (15 mm) ID. The meter couplings associated with this pipe continued to use ¾-in. straight threads on the nuts and used ½-in. iron-pipe external tapered threads on the tailpiece.

In 1896, one manufacturer, at a customer's request, built a ⅝-in. (15-mm) meter with connections for ¾-in. (20-mm) pipe. The sizes of the threads and couplings previously used on the full ¾-in. (20-mm) meter were used. The meter body then had 1-in. (25-mm) threads with ¾-in. (20-mm) couplings and was called the ⅝-in. (15-mm) × ¾-in. (20-mm) meter. This designation has become universal.

At present, threads on meter bodies having external-threaded ends are typically one nominal iron-pipe size larger than the nominal size of the meter.

Plastic Parts and Components

Several of the major meter manufacturers started experimenting with the use of plastic materials in water meters around 1960. Engineered plastics were first used as a replacement for internal parts such as measuring chambers and then expanded to other parts such as the register. By 1970, most of the manufacturers had converted to plastics for certain parts in their water meters. By the mid-1970s, a number of utilities were using meters with plastic main cases.

Plastic main-case meters require careful handling in the field during insertion into the meter setting to prevent cross threading or possible damage caused by misaligned meter settings. In the absence of a metal meter yoke, an electrical ground strap should be installed around either plastic or metal main-case meters in compliance with the National Electrical Code. Register lenses of plastic material are more easily scratched than glass and can obscure the register face if care is not exercised by meter readers when wiping the lens clean.

Water Meter Useful Life

The useful life of a water meter can be highly dependent on water quality and consumption patterns over the course of both time-in-service and volume throughput. It will also depend on the make, model, and metering technology being employed. It can also be a subjective decision, based upon factors such as water rates, water accountability concerns, and the costs of meter replacements. Manufacturer warranty periods can be one criterion in assessing useful meter life.

Remote Reading: Visual, Semiautomated, Automatic Meter Reading (AMR) and Advanced Metering Infrastructure (AMI)

This manual focuses on water meters and the measurement of water and only touches on the method for reading and transmitting data and information from water meters. Many utilities are choosing to install AMR, which is commonly referred to as either "mobile AMR" or "drive-by AMR." AMI is commonly referred to as "fixed-network AMI" or variant reading systems, such as hybrid AMR/AMI networks and cellular networks.

The first known remote-reading device, patented in 1890 by L. Ehrlich (patent 428,900), was an electrically operated unit for gas meters. A water meter remote reader, also electric, was patented in 1917 by Edwin H. Ford and A. Neff (patent 1,244,634). In 1957, the first of the modern remote readers was introduced for quantity sales. It consisted of an electric pulse generator driven by the meter with the pulses actuating a solenoid-operated counter at the remote location (patent 3,118,075 issued to M.J. Dunn).

In 1959, a meter with a battery-powered readout was introduced. When it was inserted into a plug socket at the remote location, the wiring to the meter was energized. The position of the hands on the meter dial was transmitted electrically to dials on the readout instrument carried by the meter reader. Another type, a pneumatic pulse transmitter that had a pulse-operated, remote counter, appeared in 1967.

In the 1980s, semiautomated reading evolved into reading via touch pad devices. Starting in the 1970s, fixed network meter reading employing different communication networks like telephone lines, cable TV, and electric power lines was piloted. In the late 1990s, a clear movement to radio drive-by systems became the most practical AMR system with hardware, software, meters, and installation and support services that later evolved to include the radio-based AMI system using fixed communication networks.

The development of AMR and AMI networks and meter data analytics is beyond the scope of this manual. Further information on present types of remote-reading meters is given in Chapter 9, ANSI/AWWA C706, Direct-Reading Remote-Registration Systems for Cold-Water Meters, and ANSI/AWWA C707, Encoder-Type Remote-Registration Systems for Cold-Water Meters.

Battery life is an important issue of consideration for both AMR and AMI system endpoints as well as static water meters because an endpoint can neither transmit to nor receive data from either the AMR or AMI system if the internal battery is not functioning. Whereas a mechanical register on a mechanical water meter may continue to provide a reading for an indefinite period of time, an electronic register on a mechanical meter may rely upon an internal battery (if not connected to an external power supply) to provide a reading. Most small and intermediate static water meters depend upon an internal battery to function, and many large static meters also rely upon an internal battery to function although external power sources are available on some types and models of large static meters.

REFERENCE

Crainic, M.S. 2012. "A Short History of Residential Water Meters Part 1 Mechanical Meters with Moving Parts." Proc. *Installation for Buildings and Ambiental Comfort Conference* XXI- edition. Timisoara, Romania.

BIBLIOGRAPHY

Omaghomi, T., and S. Buchberger. 2014. "Estimating Water Demands in Buildings." *Procedia Engineering*, 89:1013–1022. doi:10.1016/j.proeng.2014.11.219.

This page intentionally blank.

Chapter 2

Selecting Meter Types

INTRODUCTION

The process of selecting water meters has become increasingly complex due to considerations related to the many installation settings, customer types and applications, and an expanding array of available metering system products. For example, selecting a water meter for a typical residential application may include the following concerns: fire sprinkler regulations set for the utility, requirements for irrigation, the impacts of water scarcity, water conservation requirements, and plumbing fixture efficiency. Properly selecting water meters for accounts with widely varying flow profiles (like those commonly experienced in both commercial and industrial installations), however, requires much additional research into how water is currently being used at the facility, a fundamental knowledge of the many different types of water meters that are available, and whether or not there is a fire-suppression system served by the water meter. As noted later in this chapter, ongoing monitoring of key customer accounts is an important consideration for water meter selection.

Selection primarily involves both the size and type of water meter, although water meter register compatibility with either existing or proposed automated meter reading (AMR) or advanced metering infrastructure (AMI) system technology is also important. Too often, water meter size is chosen merely to match the supply pipe size, but oversized pipe is often installed to allow for possible future increases in water use or to reduce pressure loss in a long length of pipe. Chapter 3 discusses the method of calculating total friction loss, although the example covers only small services. Based upon this principle, plus customer demand information and pressure-loss data from water meter manufacturers' specification sheets, the meter size is selected.

The responsible party should also confirm that the proposed size and type of meter is approved by the plumbing codes that are applicable to that particular jurisdiction. For example, if a 3-in. (80-mm) meter in a 4-in. (100-mm) supply line is appropriate, use of a 4-in. (100-mm) meter is inadvisable. The oversized water meter would be unnecessarily expensive, heavier to handle, and less accurate under certain usage conditions. Depending upon the type and model of water meter selected, it may also be much more expensive to maintain. It is advisable to provide or require additional space in the meter setting, whether in a pit/vault or in an inside location, to allow for changes in meter sizes, meter types, testing spools/tees, and strainers, as usage demands and profiles change. Utilities might also need to consider the impacts that rate structures and fee schedules can have upon meter selection decisions. Service charges and fees that are based upon meter sizes could influence customers to request changes in the size of meter serving the account, provided there are minimal impacts on meeting their usage patterns and demands.

The type of meter to use is determined by the anticipated range of flow rates, plus allowable pressure loss and possible safety requirements (such as fire-service regulations). Several types are available, as shown in Table 2-1. This manual describes only the more common types; information on other meters can be obtained from the respective manufacturers. When calculating flow demand, it is important to consider meter type and suitability for use as a revenue-collecting metering device or a flow-monitoring device. This manual does not address flowmeters larger than 72 in. Large diameter flowmeter applications are covered by AWWA Manual M33, *Flowmeters in Water Supply*.

Meter sizing and selection activities should be considered an ongoing process. Though the initial selection of meter size and type is mostly driven by customer-provided data, the utility needs to be proactive regarding meter selection decisions after the initial installation. For example, usage profile information should be collected for top users to validate initial decisions or used to help make selection decisions for new accounts that might have similar demand patterns. Top users should also be monitored for significant changes in usage to determine if meter size or type changes need to be made at the account. Similar monitoring should be considered when accounts change ownership.

Water quality should also be taken into consideration when selecting a meter type. Examples of water quality issues that may affect meter performance of mechanical meters include hard water (defined by the World Health Organization in WHO/HSE/WSH/10.01/10/Rev/1 *Hardness in Drinking-water* as water with greater than 180 ppm $CaCO_3$), sand, and other debris (often referred to as foreign object debris [FOD]). Soft waters with high bicarbonate alkalinity (greater than 240 mg/L) and/or with high concentrations of both sulfate (greater than 12 mg/L) and chloride (greater than 64.9 mg/L) may cause tuberculation, but tuberculation only occurs with either bronze, cast-iron, or ductile-iron meter bodies or flow tubes. Tuberculation does not occur with stainless steel or polymer composite meter bodies or flow tubes. High concentrations of sequestration agents (polysulfates and orthosulfates) may adversely affect accuracy regardless of the meter type.

Topics to consider when selecting a meter type and meter size include but are not limited to the following:

- water quality,
- expected flow range,
- government regulations,
- installation requirements,
- long-term accuracy requirements,
- maintenance costs,
- testing costs,
- warranty terms,
- fire-service needs,
- AMR/AMI system compatibility,
- head loss, and
- local plumbing codes.

Table 2-1 Types of water meters used by water utilities

Type	Usual Range of Sizes *in.*	*mm*
I. Displacement meters		
A. Nutating disc	½–2	13–50
B. Oscillating piston	½–2	13–50
II. Nondisplacement meters		
A. Velocity		
Multijet	⅝–2	15–50
Singlejet	⅝–6	15–150
Fluidic-oscillator	½–2	13–50
Vertical-shaft type (Class I turbine)	¾–6	20–150
In-line high-velocity type (Class II turbine)	1½–20	40–500
Propeller	2–72	50–1,800
Proportional	3–12	80–300
B. Differential pressure		
Fixed opening, variable differential		
Orifice	> 1	> 25
Venturi, flow nozzle, flow tube	1–120	25–3,000
Pitot tube	Unlimited	Unlimited
Variable opening, fixed differential rotameter	¼–4	6–100
C. Electronic velocity		
Electromagnetic and ultrasonic (Type I)	½–8	12–200
Electromagnetic and ultrasonic (Type II)	½–20	12–500
D. Level measurement		
Weir, Parshall flume, etc.	Unlimited	Unlimited
III. Compound		
A. Standard compound	2–10	50–250
B. Fire-service Type I and Type II	3–10	80–250
C. Fire-service Type III	3–10	80–250
IV. Residential fire sprinkler systems in one- and two-family dwellings and manufactured homes	¾–2	20–50

METERS (1 IN. AND SMALLER)

In the United States and Canada, the standard meter for 1-in. (25-mm) and smaller sizes has typically been the displacement meter, of which there are two variations: the nutating piston (disc) and the oscillating piston. Essentially equal in performance, experience has shown that these meters are unrivaled for their combination of accuracy, long life, simple design, moderate cost, and easy maintenance. Tens of millions of these meters are in use. Displacement type meters are covered in ANSI/AWWA standards C700, Cold-Water Meters—Displacement Type, Metal Alloy Main Case, and ANSI/AWWA C710, Cold-Water Meters—Displacement Type, Plastic Main Case. Since the 1960s, multijet meters have been successfully used in the same applications. Multijet meters are covered in ANSI/AWWA C708, Cold-Water Meters—Multijet Type.

In the 1990s, singlejet technology was introduced from Europe. Singlejet meters have since been in use for small meter applications. These meters are covered in ANSI/AWWA C712, Cold-Water Meters—Singlejet Type, the first edition of which was approved in 2002.

In addition, major advancements have been made in the area of small electronic meters designed primarily for residential service. These include the fluidic-oscillator, electromagnetic, and ultrasonic meters. The standard for fluidic-oscillator meters is ANSI/AWWA C713, Cold-Water Meters—Fluidic-Oscillator Type, and the standard for both electromagnetic and ultrasonic meters is ANSI/AWWA C715, Cold Water Meters—Electromagnetic and Ultrasonic Type, for Revenue Applications.

Most residential services are metered with the ⅝-in. (15-mm) size having ¾-in. (20-mm) connections, although some states and local jurisdictions require a ¾-in (20-mm) size having 1-in (25-mm) connections and some require a 1-in. (25-mm) size having 1¼-in. (32-mm) connections. For information on flow rate in relation to pressure loss, see Chapter 3. In the absence of the meter manufacturer's recommendations regarding maximum intermittent and maximum continuous flows, positive displacement-type meters should not be run for extended, recurring periods of time at flows greater than approximately one-half of the upper value for the normal test flow limits if long life is the objective. For multijet and singlejet type meters, in the absence of meter supplier normal and high normal flow guarantees, normal flows should not be more than approximately three-fourths of the high normal flow limit if long life is the objective. Normal flows may be at the high normal flow rate limit for fluidic-oscillator, electromagnetic, and ultrasonic type water meters without adversely affecting the service life of the meter. As flow rate increases, however, the incidences of pressure transients (water hammer), possible fatigue, and localized cavitation should be taken into sizing consideration.

Utilities have some flexibility and options to consider in cost effectively changing 1 in. and smaller meter sizes depending upon the meter connection and customer demand situation. For example, ⅝-in. meters can be upsized with a ¾-in. short meter. A ¾-in. or 1-in. meter can be replaced with a smaller size meter with the use of the appropriate hardware. Upsizing existing ⅝-in. and ¾-in. meters to a 1-in. size is more costly, as there might not be sufficient space within the existing meter setting for a larger size meter. However, in such situations, using an electronic meter with a wider flow range and no mechanical wear is an option to consider.

INTERMEDIATE METERS (1½ AND 2 IN.)

For services that can be metered with either 1½-in. (40-mm) or 2-in (50-mm) meters, the displacement, multijet, singlejet, ultrasonic, or electromagnetic type meter can be used. Common applications for this size of meter include large residences with lawn irrigation systems, small businesses, professional offices, municipal buildings, and small fast-food restaurants. Certain models of axial flow turbine meters whose performance characteristics exceed the minimum standards set forth in ANSI/AWWA C701 Cold-Water Meters–Turbine Type, for Customer Service may also be used in this application. The low-flow accuracy of modern 1½-in. (40-mm) meters is excellent, and compound meters are not usually manufactured in sizes smaller than 2 in. (50 mm). Where low-flow accuracy is not important, as in services used only for filling tanks at rapid flow or for the irrigation of larger lots, the 1½-in. (40-mm) Class II turbine meter may be considered.

While technically classified as large meters, 2-in. (50-mm) compound meters are designed to accurately measure usage across a wide range of flows from very low to high. Most 1½-in. (40-mm) and 2-in (50-mm) meters Class II turbine type meters are designed to accurately measure usage across both intermediate and high flow rates. As stated above, certain models of axial flow turbine meters whose performance characteristics exceed the minimum standards set forth in ANSI/AWWA C701 may also be used in these applications because of their ability to accurately measure usage at low, intermediate, and high flow rates.

Those who prefer the positive displacement meter have indicated that it is simpler to use, lower in cost, and records consumption remarkably well at low flow rates. The life expectancy of any positive displacement, multijet, singlejet, or turbine type meter may be shortened if exposed to excessive FOD. Proponents of compound, ultrasonic, and electromagnetic meters emphasize that these meter types are more accurate at very low flow rates, such as when there is a leak at the property, and these meter types have lower pressure loss at high flow rates. Proponents of multijet, singlejet, ultrasonic, and electromagnetic meters also emphasize the quiet operation and the ability of these meter types to pass small particles and other entrained flow FOD. The presence of FOD seems to have a negligible effect on the life expectancy of those types of meters, although larger FOD may clog a strainer element if one is present.

Initial procurement cost is not the only consideration for determining meter selection. Installation (and reinstallation), long-term accuracy, maintenance and testing costs, and warranty coverage also need to be considered. The length of time for accuracy warranties for larger mechanical meters is typically much less than that for 1-in. and smaller meters, requiring more frequent testing to determine when accuracy starts to degrade to better determine replacement and repair strategies. Electronic meters normally have longer accuracy warranties and have no serviceable parts. However, water quality issues could impact accuracy, and frequency of testing these meter types to flag various issues is not well known at this time.

The choice between compound, singlejet, ultrasonic, electromagnetic, or turbine meters appears to be determined by the expected flow rates to be measured and the usage profile of the customer. Significant and periodic usage at different flow rates need to be compared to the manufacturers' technical information to achieve the best measurement technology fit. If accuracy at very low flow rates is important, but large capacity is also needed, the compound, singlejet, ultrasonic, electromagnetic meters, or certain models of axial flow Class II turbine meters whose performance characteristics exceed the minimum standards set forth in ANSI/AWWA C701 should be used. If large capacity is of primary importance because the typical flows are usually above 10 percent of the low normal flow rating for the meter, the turbine meter may be used.

ANSI/AWWA C714 Cold-Water Meters for Residential Fire Sprinkler Systems in One- and Two-Family Dwellings and Manufactured Homes covers the minimum requirements for cold-water meters for residential fire sprinkler applications that meet the requirements of National Fire Protection Association (NFPA) 13D in single- and two-family dwellings and manufactured homes, in sizes ¾ in. (20 mm) through 2 in. (50 mm), including materials and design. Subjects UL 327A and UL 327B cover UL Standards & Engagement (ULSE) requirements for residential fire sprinkler meters for residential structures.

LARGE METERS (3 IN. AND LARGER)

The water utility now has many choices of water meters for lines 3 in. and larger: compound, turbine, propeller, singlejet, ultrasonic, electromagnetic, or fire-service meters, as well as certain models of axial flow turbine meters with performance characteristics that exceed the minimum standards set forth in ANSI/AWWA C701. If there is a fire-suppression system downstream of the water meter, it is advisable to choose a meter that either meets or exceeds ANSI/AWWA C703 Cold-Water Meters—Fire-Service Type and is either ULSE listed or FM approved for fire-service use. Ultrasonic or electromagnetic meters (see ANSI/AWWA C715) provide an advantage over traditional compound and turbine type meters for both commercial and industrial applications. Regardless of the type of large water meter selected, however, it is imperative that the meter be installed according to the manufacturer's installation instructions. Any deviation from these guidelines may adversely affect both meter accuracy and useful life.

The ability of a meter to accurately measure usage at low flow rates is usually very important at places in which people live (e.g., condominiums, apartment complexes, assisted living centers, hotels), shop, work, and attend either primary or secondary school. Nursing homes, hospitals, hotels, and condominiums or apartment complexes with elevated water storage tanks that have large setpoint differences between the level at the end of one refill cycle and the (lower) level at the start of the next refill cycle, however, typically have only minimal usage at low flow rates, and turbine meters may be used for these services. It is advisable for a utility to periodically review all of its large meter services to make certain that the correct type of meter is installed for each application, as customer usage patterns may change over time.

The traditional compound meter registers well over a wide flow range and has relatively low pressure loss at high flow rates. There are three basic types of compound meter:

- single body and single (totalizing) register,
- single body and two registers (one register for the bypass meter and one register for the main meter), and
- dual body and two registers.

For compound meters with two registers, it is important that the billing department knows that each register needs to be billed.

All compound meters lose a certain degree of accuracy operating within the changeover flow range by varying amounts. It is important to size any compound meter installation to minimize the total flow each installation experiences in the changeover flow range. To help ensure that the proper size of a compound meter is selected or even if it is the right measurement technology type of the application, the utility will need to collect meter accuracy information through the changeover range and flow profile information to determine the amount of usage that occurs within the changeover range.

Turbine meters are used when flows are large, minimum flows are usually above 10 or 12 percent of maximum rating, and the low-pressure loss at high flows is valuable. These meters have a separate inner measuring chamber or cage, which may be removed with the rotor for servicing. ANSI/AWWA C701 divides turbine meters into Class I and Class II meters. Both classes of meters register by recording the revolutions of a turbine set in motion by the force of flowing water striking the blades. Class I turbine meters are vertical-shaft models, while Class II turbine meters are in-line (high-velocity) models. The former extend to somewhat lower low-flow capabilities, while the latter have somewhat higher high-flow capabilities and lower head loss.

For nonrevenue applications, ANSI/AWWA C750 Transit-Time Flowmeters in Full Closed Conduits, addresses transit-time ultrasonic flowmeters used in a variety of applications as flow-monitoring devices (rate measurement). Magnetic flowmeters for nonrevenue applications are covered in ANSI/AWWA C751 Magnetic Inductive Flowmeters.

Propeller meters are intended for mainline or pump station discharge measurements. They operate with the minimum of pressure loss for a mechanical meter. In small propeller meters, the propeller is built into a section of pipe of precise diameter and forms part of the meter; in large sizes, saddle-mounting of the propeller is customary, with the interior of the pipe serving as the measuring tube. At the point of measurement, the velocity profile should be uniform. This requires that there are no elbows, valves, constrictions, or enlargements in the pipe for several pipe diameters of distance upstream and for a lesser distance downstream. For detailed instructions, users should consult the manufacturer of the proposed meter.

Singlejet meters can accurately measure usage across a very wide range of flows. Following the manufacturer's installation instructions is critical for singlejet meter accuracy and to achieve the design life of the meter. It is always a best practice to install a strainer upstream of a singlejet meter to condition the flow and to protect the impeller from damage by FOD.

Advances in battery technology as well as in computing technology have made both ultrasonic and electromagnetic meters common for revenue applications, as well as for raw and finished water applications, industrial processes, and wastewater applications. Both meter types offer the advantage of a having either an unobstructed or a nearly unobstructed flow tube with no moving parts, and both provide high accuracy across a wide range of flows. Excluding the power source or battery, there is nothing on these two meter types to be maintained. The installer should carefully follow the manufacturer's installation instructions for both ultrasonic and electromagnetic type meters, especially in regards to lengths of straight pipe both upstream and downstream of the meter. Electromagnetic and ultrasonic meter accuracy may be compromised if the flow is not conditioned properly. Consult with the meter manufacturer about whether the installation of a strainer might improve the conditioning of the flow profile. Lastly, some electromagnetic meters may require electrical grounding.

Many models of large meters have epoxy coated, gray iron, ductile iron, or carbon steel bodies. It is recommended that the installer use stainless steel flat washers on both the underside of the bolt head and against the nut to protect the epoxy coating that serves to prevent rust from forming on the meter body.

Where fire hydrants or large sprinkler systems are served by the metered line, insurance underwriters usually insist upon using a fire-service meter. Where these service conditions are encountered, the applicable insurance requirements should be ascertained before the meter is selected. Fire-service meters permit day-to-day usage to be drawn from a line serving hydrants or sprinklers, with the entire flow metered.

ANSI/AWWA C714 and ULSE Subjects UL 327A and UL 327B cover NFPA 13D applications. ANSI/AWWA C703 and both the UL 327 Standard and the FM 1044 Standard generally address larger NFPA 13 Systems. The person selecting fire-service meters should be aware of a "gap" in coverage by the existing standards: certain NFPA 13R systems may require fire-service meters smaller than those covered in ANSI/AWWA C703, but ANSI/AWWA C714 specifically excludes NFPA 13R systems.

For detecting the unauthorized use of water in lines for unmetered fire-service applications, a unit called a detector check, double check detector assembly, or reduced pressure principle detector assembly may be installed. This same construction permits authorized metered use of small quantities of water from a fire-service line. The detector check meter, a ULSE-listed, FM-approved device, is designed for dedicated fire-service applications. These devices are not intended to meter consumption; rather, utilities use these devices to detect leaks and unauthorized water usage. A second, normally smaller, service line is installed with an appropriately sized water meter for measurement of normal domestic water consumption. The detector check meter is an appropriate device for many commercial and industrial fire systems. If, however, the water is from an outside source or additives are present for antifreeze, regulatory agencies for fire-protection systems require a testable backflow preventer. These double check or reduced pressure assemblies are also available with metered bypasses. AWWA Manual M14, *Backflow Prevention and Cross-Connection Control,* contains more information.

Other types of meters for large lines include the Venturi or its modifications, the flow tube, or the flow nozzle. These meters differ from the other meters because they measure by differential pressure instead of by velocity or quantity. Flow ranges for accuracy vary with size and with details of design, so information should be obtained from the manufacturer of the meter. Although built in smaller sizes, these meters usually are used in 6-in. (150-mm) or larger lines.

For temporary measurements, the Pitot tube may be used in main lines. In the hands of skilled operators, the meter is valuable in making surveys to ascertain rates of flow in large lines, to locate leaks, and to measure the flow from fire hydrants.

Manifolding of Meters

For 4-in. (100-mm) or larger lines, instead of using one meter to measure the entire flow, multiple, smaller-sized meters may be installed in a manifold (also called a battery). Typically, a manifold should consist of two or more meters, with each meter one pipe size or more smaller than the main line. On a 6-in. (150-mm) or 10-in. (250-mm) line, the manifold may consist of two or three compound, singlejet, ultrasonic, electromagnetic, or turbine meters. It is advisable that all meters in a manifold should be of the same brand, model (type), and size.

Branch manifolds are only recommended if there is close coordination of the technical details with the meter manufacturer or other experts in flow hydraulics. It is imperative that the flow through the multiple branches be balanced for proper meter registration. If the flow is not balanced properly, one of the meters in the manifold assembly could experience reverse flow registration or the whole assembly suffer reduced performance. Strainers serving meters installed in a manifold configuration also need periodic inspection and cleaning to ensure there is no FOD buildup affecting the flow profile to the meter or disrupt to flow to other meters within the manifold. Figure 4-5 in Chapter 4 illustrates a typical 3-meter and 2-meter hydraulically balanced manifold of large meters.

Manifolding of meters has the following advantages.

- One meter can be serviced by closing its shutoff valves while the water supply is still available to the customer through the other meter.
- Since the meters within the manifold are smaller than a single meter for the same service, fewer personnel may be required for meter servicing or removal, but this is not necessarily true with the meters in a larger manifold.
- Inventory of repair parts can be minimized, and the knowledge of repair and testing techniques can be less involved when the variety of size and type of meters in the system is controlled.
- The manifold can be side-wall mounted to conserve floor space.
- Two or possibly three smaller meters may cost less to purchase than one large meter, but this is not always the case. Moreover, certain types of large meters with electronics that are battery powered either may need to be replaced or have their measuring chamber assembly replaced after a decade of service, whereas some models of intermediate meters may remain in service for two decades. In addition to a hydraulic analysis, the designer should perform an economic analysis on choosing a single large meter or a manifold of smaller meters. The cost of installing multiple meters, when the costs of inlet and outlet valves, required fittings (i.e., to create the manifold assembly), piping, labor, and AMR/AMI system endpoints (if required) are included, may exceed the cost of a single large meter.

Chapter 3

Effects of Water-Flow Friction, Increased Usage, and Pressure Losses on Service Adequacy

INTRODUCTION

Careful consideration should be given to a service-line installation to ensure that it will provide an adequate supply of water to a customer. Frequently, new water services are sized and installations made based on experience. However, although experience does contribute considerably to a satisfactory and adequate supply, other important factors must be evaluated. It is possible to have adjacent installations that vary by peak demand to such an extent that they are not identical. Each installation should be considered individually and the proper size installation made initially. After startup, changes are expensive and often impossible to make.

FACTORS TO CONSIDER IN SIZING MAINS

The following are important factors to consider in sizing mains:

- water demand and expected usage,
- pressure and volume of water presently available and in the future at the point-of-service connection in the main,
- projected future installations,
- pressure loss due to differences in elevation between the main and building,
- water-flow friction in the service line from the main in the street to the curb or property line,
- water-flow friction in the service line from the curb or property line to the building,
- pressure loss through the meter and valves, and
- pressure loss and water-flow friction in the building.

The utility is responsible for providing an adequate supply of water at sufficient pressure to its customers. In addition to meeting present or initial demands, anticipated usage should be considered as well. Mains are often extended to accommodate new housing developments. This increased water usage could cause less water to be available and at lower pressure. Therefore, the service installation to a building should be slightly larger under these circumstances to compensate for the expected supply variance.

There must be sufficient pressure in the main to elevate the water to the building. It is obvious that if the installation is high above the main, for example on a hill, more pressure is required to deliver the water than if it were in a valley below. One psi of pressure will raise water 27.7 in. or 2.31 ft (0.704 m) in a vertical pipe. This is called elevation head, and the total difference in elevation head of an installation must be subtracted from the available main pressure when considering available operating pressure. When the dwelling is below the main (Table 3-1), the available operating pressure increases and the elevation head may be added to main pressure (at the rate of 1 psi [6.9 kPa] for every 2.31 ft [0.704 m] in height). The distance from the main to the property line may vary in each installation, especially when the main is placed along one side of the street and service must be made to both.

Table 3-1 Elevation above supply main: elevation head

Elevation to Building, *ft*	Head, *psi*	Elevation to Building, *ft*	Head, *psi*
1	0.433	14	6.062
2	0.866	15	6.495
3	1.299	16	6.928
4	1.732	17	7.361
5	2.165	18	7.794
6	2.598	19	8.227
7	3.031	20	8.660
8	3.464	21	9.093
9	3.897	22	9.526
10	4.330	23	9.959
11	4.763	24	10.392
12	5.196	25	10.825
13	5.629		

NOTES:
Other values of elevation head can be calculated by multiplying actual elevation, in feet, by 0.433 (head at 1 ft).
Metric conversions: ft × 0.3048 = m, psi × 6.89476 = kPa.

Friction loss due to water flow through the service line from the property line to the building must also be considered. The differences not only between tubing of various sizes but between copper and plastic tubing are illustrated in Table 3-2. Both disc and oscillating-piston meters, if they are to meet AWWA standards, must be manufactured to the stated material and performance requirements. Consequently, both have comparable pressure-loss features. The pressure losses shown in Table 3-3 are averages and were computed using a 10-psi (69-kPa) pressure loss at maximum flow for all-size meters. This is an ample figure for estimating flow and is applicable to both meters. Actual figures can be obtained from the manufacturer of the meter and used in the calculations. It is difficult, and of little importance, to show in detail the differences among all meters manufactured because the difference is slight when the meters are of similar size.

As a general rule, the water utility does not have the responsibility for or control over the pipe size and fittings within a customer's building, so allowance must be made for reasonable customer pressure losses and adequate flow provided. Automatic washing machines, dishwashers, garbage disposals, and multiple bathrooms have created the need for more flow-rate capacity; consequently, service requirements have increased considerably. In order to provide adequate water-supply pressure, losses must be minimized in the service installation.

FRICTION-LOSS DATA

Table 3-2 has been compiled from computed data to illustrate the relationship and comparative pressure loss between copper tubing, plastic tubing made to iron-pipe inner-diameter (ID) dimensions, and plastic tubing made to copper-tube outer-diameter dimensions. It should be noted that no reference has been made to data for ½-in. (13-mm) copper or plastic service tubing, nor has the table been included. This omission was deliberate to discourage the use of this size. Pressure loss is considerably higher (approximately five times at 10 gal per min [gpm] [2.3 m^3/h]) through ½-in. (13-mm) copper tubing when compared with the ¾-in. (20-mm) size. Since the difference in initial material cost is negligible with identical installation, ½-in. (13-mm) service lines are not recommended.

Careful review of Table 3-2 will disclose that data are shown in pounds per square inch per linear foot. This enables the user to multiply the length of service line by the factor shown for the required flow rate. As an illustration, the water-friction loss through 88 ft (26.8 m) of ¾-in. (20-mm) copper tubing at 15 gpm (3.4 m^3/h) is $0.263 \times 88 = 23.14$ psi (159.5 kPa), whereas the loss through ¾-in. (20-mm) plastic tubing, iron-pipe ID size, is $0.169 \times 88 = 14.87$ psi (102.5 kPa).

SAMPLE SERVICE INSTALLATION CALCULATION

When an actual installation is calculated, all necessary facts should be available. For the purpose of illustration, a hypothetical service will be considered. The water-supply main is 5 ft (1.5 m) below the pavement. The length of the service line from the water main to the property line is 43 ft (13.1 m) and 88 ft (26.8 m) from the property line to the meter. The meter is to be installed in the basement of a two-story dwelling, 21 ft (6.4 m) above the pavement. The main water pressure available at the corporation connection is 63 psi (435 kPa). Assume a household flow-rate requirement of 15 gpm (3.4 m^3/h) and 30-psi

Table 3-2 Comparison of computed water flow friction losses for service line, *psi/ft*

	¾ in.			1 in.			1¼ in.	
Flow Rate *gpm*	Copper Tube	Plastic* Iron Pipe ID	Plastic Copper Tube OD	Copper Tube	Plastic* Iron Pipe ID	Plastic Copper Tube OD	Copper Tube	Plastic* Iron Pipe ID
1	0.003	0.002	0.005					
2	0.008	0.005	0.013					
3	0.016	0.010	0.026					
4	0.025	0.015	0.041					
5	0.037	0.024	0.065	0.009	0.007	0.022		
6	0.048	0.031	0.081	0.011	0.008	0.029		
7	0.063	0.041	0.108	0.018	0.011	0.038		
8	0.075	0.049	0.128	0.021	0.013	0.045		
9	0.095	0.062	0.160	0.026	0.017	0.057		
10	0.121	0.078	0.203	0.035	0.021	0.072		
11	0.166	0.108	0.274	0.057	0.040	0.104		
12	0.192	0.128	0.317	0.069	0.048	0.126		
13	0.206	0.134	0.338	0.071	0.050	0.130		
14	0.228	0.149	0.377	0.086	0.059	0.156		
15	0.263	0.169	0.428	0.107	0.076	0.200		
20	0.433	0.286	0.734	0.120	0.085	0.225	0.042	0.030
25	0.656	0.429	1.118	0.226	0.136	0.343	0.059	0.046
30	0.920	0.599		0.242	0.188	0.450	0.083	0.060
35		0.809		0.370	0.254	0.611	0.108	0.076
40				0.411	0.317	0.763	0.135	0.091
45				0.517	0.397	0.924	0.156	0.109
50				0.623	0.482	1.100	0.200	0.131
55				0.749	0.583		0.234	0.161
60				0.892	0.685		0.278	0.185

* Plastic per AWWA C901, Tables 2 and 3.
NOTE: Metric conversions: gpm × 0.2268 = m^3/h, in. × 25.4 = mm, psi/ft × 22.62 = kPa/m.
gpm = gallons per minute, ID = inner diameter, OD = outer diameter

(210-kPa) pressure. This rate will vary depending on the number of residents and the household water-using equipment. (See Table 3-4 for theoretical discharge from various fixtures.) Actual discharge is usually less due to rough casting surface, etc. If pressure at the fixtures is above approximately 80 psi (550 kPa), a pressure-reducing valve should be installed in the system to avoid excessive force of water flowing from the fixtures.

Because the water-supply main is 5 ft (1.5 m) below the pavement, and the meter is 21 ft (6.4 m) higher than the pavement, the total elevation head to the meter is 26 ft (7.9 m). Table 3-1 is shown only to 25 ft (7.6 m), but the total elevation head can be calculated by multiplying 26 × 0.433, which equals 11.26 psi (77.6 kPa) (see Table 3-1 footnote). This figure must be added to the 30-psi (210-kPa) operating pressure required to service the fixtures and faucets in the building in order to obtain the operating pressure required at the main. A ¾-in. (20-mm) service will be considered and calculated. If a ¾-in. (20-mm) corporation stop is installed (Table 3-3), the pressure loss will be 1.70 psi (11.7 kPa). The total water-flow friction must now be calculated by adding the length of tubing from main

Table 3-3 Computed pressure losses for service components, *psi*

	Corporation Stop		Curb Stop		Globe Valve		Gate or Ball Valve		Mechanical Meter			Static Meter Type I & II		
Flow Rate	3/4*	1†	3/4‡	1§	3/4**	1††	3/4‡‡	1§§	5/8	3/4	1	5/8	3/4	1
gpm	*in.*		*in.*		*in.*		*in.*		*in.*					
1	0.015	0.004	0.011	0.003	0.060	0.017						0.31	0.29	0,25
2	0.050	0.014	0.034	0.008	0.195	0.058	0.001					0.46	0.39	.31
3	0.103	0.029	0.071	0.017	0.301	0.09	0.002					0.65	0.51	0.38
4	0.170	0.03	0.117	0.028	0.664	0.20	0.003		0.39			0.88	0.65	0.45
5	0.246	0.07	0.170	0.04	0.962	0.29	0.005	0.002	0.62			1.15	0.81	0.52
6	0.340	0.11	0.234	0.05	1.33	0.43	0.007	0.002	0.89	0.39		1.46	0.98	0.61
7	0.457	0.14	0.315	0.07	1.79	0.57	0.009	0.004	1.22	0.53		1.81	1.16	0.69
8	0.557	0.17	0.384	0.08	2.18	0.71	0.011	0.004	1.60	0.71	0.25	2.20	1.36	0.79
9	0.703	0.22	0.485	0.10	2.75	0.85	0.013	0.005	2.02	0.89	0.32	2.63	1.58	0.89
10	0.879	0.23	0.606	0.13	3.44	1.0	0.016	0.007	2.50	1.10	0.39	3.10	1.82	1.00
11	1.20	0.28	0.72	0.15	4.08	1.2	0.020	0.009	3.00	1.35	0.48	3.61	2.07	1.11
12	1.31	0.32	0.83	0.18	4.72	1.4	0.024	0.011	3.60	1.60	0.57	4.16	2.34	1.22
13	1.40	0.37	0.94	0.21	5.36	1.6	0.028	0.013	4.20	1.88	0.67	4.75	2.63	1.35
14	1.54	0.42	1.05	0.24	6.00	1.8	0.032	0.015	4.90	2.18	0.77	5.38	2.93	1.48
15	1.70	0.47	1.17	0.27	6.64	2.0	0.036	0.017	5.60	2.50	0.89	6,05	3.25	1.61
16	2.11	0.53	1.31	0.30	7.46	2.3	0.040	0.019	6.40	2.85	1.02	6.76	3.58	1.76
17	2.25	0.59	1.45	0.34	8.28	2.6	0.044	0.021	7.20	3.22	1.16	7.51	3.93	1.90
18	2.36	0.65	1.60	0.38	9.10	2.9	0.049	0.023	8.10	3.60	1.30	8.30	4.30	2.06
19	2.54	0.72	1.75	0.42	9.93	3.2	0.054	0.025	9.00	4.00	1.45	9.13	4.68	2.21
20	2.75	0.80	1.90	0.46	10.76	3.5	0.059	0.028	10.00	4.45	1.60	10.00	5.08	2.38
25		1.2		0.67		5.1		0.047		7.10	2.50		7.33	3.29
30		1.6		0.92		7.0		0.062		10.00	3.60		10.00	4.36
35		2.1		1.2		9.0		0.077			4.80			5.56
40		2.7		1.5		11.6		0.086			6.40			6.92
45		3.3		1.9		14.6		0.093			8.10			8.42
50		4.0		2.3		17.6		0.130			10.00			10.00

* Based on equivalent loss of 5.86 ft of 3/4-in. copper tubing.
† Based on equivalent loss of 6.67 ft of 1-in. copper tubing.
‡ Based on equivalent loss of 4.04 ft of 3/4-in. copper tubing.
§ Based on equivalent loss of 3.85 ft of 1-in. copper tubing.
** Based on equivalent loss of 22.90 ft of 3/4-in. copper tubing.
†† Based on equivalent loss of 29.1 ft of 1-in. copper tubing.
‡‡ Based on equivalent loss of 0.14 ft of 3/4-in. copper tubing.
§§ Based on equivalent loss of 0.21 ft of 1-in. copper tubing.
NOTE: Metric conversions: gpm × 0.2268 = m^3/h, in. × 25.4 = mm, psi × 6.89476 = kPa.
gpm = gallons per minute

to property line, 43 ft (13.1 m), and the distance the meter is to the property line, 88 ft (26.8 m), for a total of 131 ft (39.9 m). Table 3-2 shows that, for each foot of ¾-in. (20-mm) copper tubing, the friction loss at 15 gpm (3.4 m^3/h) is 0.263 psi (1.81 kPa). Consequently, when this figure is multiplied by the total of 131 ft (39.9 m), the total friction loss in the tubing is 34.45 psi (237.5 kPa). The total pressure required at the main is 77.41 psi (533.7 kPa). Based on these data, the system must be redesigned because only 63 psi (434.4 kPa) is available at the main and the meter and valve losses has not been considered.

In recalculation, the elevation head will not change, even though the supply lines are increased in size (elevation head is constant and irrespective of diameter). Consequently, it remains 11.26 psi (77.6 kPa). The water-flow friction through 131 ft (39.9 m) of 1-in. (25-mm) copper tubing is 14.01 psi (96.6 kPa); through a corporation stop 1 in. (25-mm) in size, the friction is 0.47 psi (3.24 kPa) and building loss is 30 psi (210 kPa) for a total of 55.74 psi (384.3 kPa). Available distribution system pressure is 63 psi (434.4 kPa). Consequently, 7.26 psi (50.0 kPa) remains for the meter shutoff valve and the meter. Table 3-3 shows that pressure loss through a ⅝-in. (15-mm) meter at 15 gpm (3.4 m^3/h) is 5.60 psi (38.6 kPa), and the 1-in. (25-mm) ball valve (or gate valve) is less than the remaining available 1.66 psi (11.4 kPa). This new system will provide sufficient water at the dwelling.

The minimum available water pressure may be used in the calculation for pressure loss. It is not necessary to have extra or reserved pressure, but it is necessary to calculate the friction and pressure properly and consider all variables. Increasing the pressure loss or flow friction by miscalculation will decrease the available flow rate, but it will not stop the flow. Flow can be stopped only by an obstruction, such as a closed valve.

When systems are designed for large meters, the same technique is used for calculating the pressure loss and flow friction. The size of the pipe, of course, will be different from that shown in the tables, and new values will be required for the flow rates. The elevation head remains constant, and only the flow friction changes, according to the flow rate and pipe size. Turbine, compound, and propeller meters have different flow characteristics, and the flow rate could vary considerably. When these meters are designed into a system, the manufacturer should be consulted for pressure loss information.

Frequently, after a service is in operation, the demand for water will increase. It is possible to increase the supply by replacing inadequate pipe or valves with material that has less pressure loss. However, it is desirable to have the correct-sized meter (as determined by the utility) installed so that maximum accuracy at minimum flow can be realized. Information discussed in this chapter is covered more extensively in AWWA Manual M22, *Sizing Water Service Lines and Meters*.

Table 3-4 Discharge from fixtures and faucets at various operating pressures

	Operating Pressure		
	5 psi	30 psi	90 psi
Type of Fixture or Faucet	Discharge, *gpm*		
¾-in. compression sink faucet:			
Wide open	8.1	20.0	33.4
Half open	7.6	19.0	32.9
One-fourth open	7.0	17.4	29.9
½-in. compression sink faucet, wide open	6.0	14.8	24.5
½-in. ground key sink faucet, wide open	9.5	23.4	36.4
¾-in. ground key sink faucet, wide open	13.8	31.7	51.0
¾-in. compression sink faucet, wide open	9.0	22.1	36.0
½-in. self-closing compression faucet, wide open	2.6	6.8	11.7
⅜-in. ground key sink faucet, wide open	6.8	16.7	27.7
⅜-in. compression sink faucet, wide open	3.2	8.2	14.1
½-in. compression sink faucet, wide open	4.8	12.3	21.3
½-in. compression laundry tray faucet, open	6.3	17.3	25.3
Compression wash basin, wide open	5.0	11.9	21.3
1-in. ground key sink faucet, wide open	30.7	78.9	118.8
1-in. compression sink faucet, wide open	12.7	39.9	64.8
Combination compression laundry faucet:			
Both outlets open	9.6	22.4	38.6
Either hot or cold, wide open	6.1	14.4	24.8
Combination compression bath tub:			
Both hot and cold open, no nozzle	8.0	20.4	34.4
Both hot and cold open, with nozzle	5.9	14.3	24.8
Either hot or cold only, open, no nozzle	4.3	11.1	19.9
Either hot or cold only, open, with nozzle	3.8	9.2	16.1
Combination compression sink faucet with swinging nozzle:			
Hot and cold open	4.6	12.2	21.4
Either hot or cold open	3.2	8.4	14.8
Water closets:			
Tank type	2.9	8.0	14.6
Flush valves	9.7	30.0	45.7

Note: Metric conversions: psi × 6.89476 = kPa, gpm × 0.2268 = m^3/h, in. × 25.4 = mm.
gpm = gallons per minute

This page intentionally blank.

Chapter 4

Meter Installation

INTRODUCTION

Water meters for customer service are installed or set either indoors or outdoors. In an indoor setting, the meter is installed inside the customer's premises, usually in the basement or a utility room. In an outdoor setting, the meter is installed underground in a pit or meter box that is usually located at the curb end of the service line. Historically, indoor settings have been used primarily in northern states where severe winter weather may cause frost damage, and outdoor settings have been used in warm, temperate climates. Now that electronic registers are more common, even some providers in temperate climates are opting for indoor settings to provide an environment that is less likely to impact the contained register electronics.

OUTDOOR VERSUS INDOOR SETTINGS

The advantages of outdoor settings include

- meter location at point of delivery to customer;
- elimination of a separate curb box;
- reduction in meter damage from water heater failure; and
- no need to enter the customer's home for reading, inspection, and replacement.

Disadvantages of outdoor settings include

- high costs when frost damage or moisture incursion occurs,
- difficulty for manual reading,
- potential for high maintenance due to flooding or snow removal,
- possible damage due to vandalism,
- possible damage from pipe misalignment,

- possible damage caused by ground movement,
- liability exposure from the public for tripping accidents, and
- possible pit modifications required due to grade changes.

The advantages of indoor settings include

- potentially lower installation costs;
- reduced damage and maintenance through elimination of exposure to outdoor conditions;
- for some utilities, the opportunity for customer contact via the meter reader; and
- the extended life of components such as automated meter reading (AMR) systems.

Disadvantages of indoor settings include

- missed manual readings because of entry problems,
- hot-water damage,
- basement flooding due to frost-protective-bottom fracture or miscellaneous leaks, and
- consumer complaints about meter readers.

Remote Registers

The advent of inexpensive and reliable AMR devices that can be read either manually or automatically has added installation capabilities that modify and, in some cases, eliminate the previously accepted advantages and disadvantages of indoor versus outdoor settings. For example, it is now possible to read meters set indoors from outside the house and eliminate entry or call-back problems. It is also possible to read outside meters installed in pits without removing meter pit lids by using remote or automatic reading devices (including some remote devices mounted in pit lids).

With the flexibility of remote registers, the utility manager can set meters to achieve minimum installation costs as well as minimum meter-reading costs. Because of the economic advantages of obtaining indoor meter readings and other metering information from outside the house, most remote-reading devices that were initially installed in northern areas are now common in the warmer climates as well. It is estimated that approximately two-thirds of new meters are shipped with some type of AMR system. Remote-reading devices are now offered by both domestic meter manufacturers and some specialty suppliers; these devices are described in Chapter 9. These devices offer an alternative in areas where outdoor pit settings have traditionally been used. The decision can now be made on the basis of overall economics, considering initial investment, installation cost, reading cost, and maintenance expense. Elimination of pit hazards, public injury liability, and pit maintenance should also be considered.

METER INSTALLATION

In addition to special considerations for indoor and outdoor meter settings, commercial, industrial, or large meter installations also have special requirements. This section includes general considerations applicable to all meter installations and their related special needs.

General Considerations

Although standard specifications exist for meters, valves, pipe, and tubing, there are no standards for meter settings; however, there are certain principles that should be observed. Specific problems and questions related to installation can best be addressed by consulting manufacturers of meters or setting and testing equipment. Over the years, a wealth of experience has been accumulated to provide meter settings that ensure optimum meter performance and service accessibility, along with ease of installation and low cost. Installation hardware is available in a wide variety of sizes, types, and materials to meet virtually any installation preference or requirement. Many cities, counties, provinces, and states have adopted standards and ordinances under building plumbing codes to cover water meter installations.

Basic requirements of an acceptable meter installation are as follows.

- Position the meter in the horizontal plane for optimum meter performance (in most sets, the on-meter register dial face should point up).
- Locate the meter so that it is readily and safely accessible for reading, servicing, and/or testing.
- Provide a leak-tight, permanent setting to ensure that the meter can be removed from service without negatively affecting customer's plumbing.
- Provide for permanent electrical grounding that does not use the meter to prevent accidental shock to meter service personnel.
- Protect the meter from freezing and other conditions that could damage the installation.
- Provide a high-quality inlet shutoff valve to allow for meter maintenance. The location of the meter may also dictate a meter valve on the outlet side to prevent water draining back when meter is removed.
- The water distribution system in which the meter is installed should have a minimal loss of pressure before the water reaches the meter.
- Consider public safety and design installations to prevent accidents.

To avoid future operating problems, all open connections should be capped whenever a meter is removed from its setting for any length of time. Meter caps should be kept on until installation to keep debris from entering into the meter, especially for outdoor settings. In the actual setting, a meter idler can be used in place of caps to provide the same protection. Also, meters should be protected from heat and direct sunlight during storage and transit before installation or after removal.

On all indoor settings, it is highly important that electrical continuity be maintained through the water line. Most utilities require electrical bonding around meters to prevent accidental electrocution of service personnel changing meters. If the meter setting itself does not provide a continuous electrical circuit when the meter is removed, a permanently bonded electrical grounding strap should be provided. Electrical grounding is a requirement specified by the National Fire Protection Association code 70, and all service and installation personnel should be advised of this safety requirement. Most commercially available prepared meter settings provide a continuous metallic circuit even when the meter is removed from the line. AWWA opposes the grounding of electrical systems to pipe systems conveying drinking water to a customer's premises. Two types of commercially available meter settings are shown in Figure 4-1.

In many North American and European countries, ample protection from backflow and backsiphonage is required in single-family dwellings, as well as at commercial and industrial sites. The water supplier should be familiar with ANSI/AWWA C510 Double Check-Valve Backflow Prevention Assembly, ANSI/AWWA C511 Reduced-Pressure Principle Backflow-Prevention Assembly, and the requirements of federal, state or provincial, and local authorities and should know which takes precedence. This ensures that proper consideration is given to the meter installation design, thereby providing for the required backflow preventers.

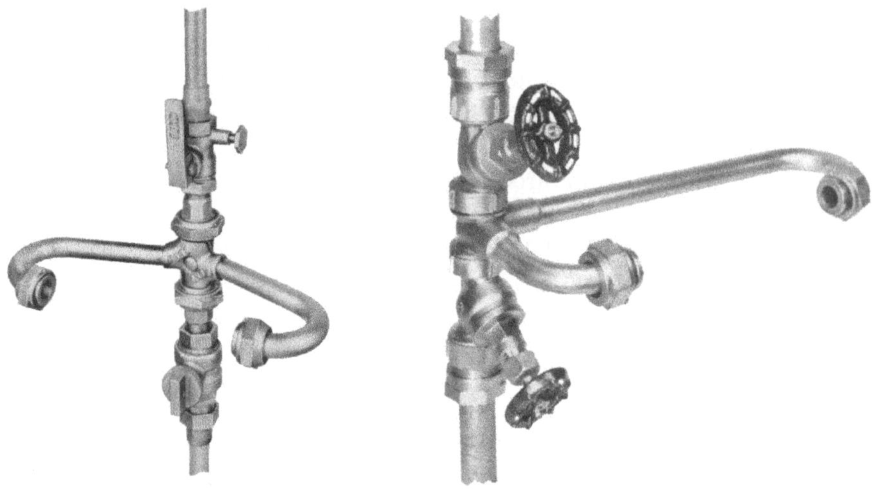

Figure 4-1 Two types of meter setters for vertical indoor piping

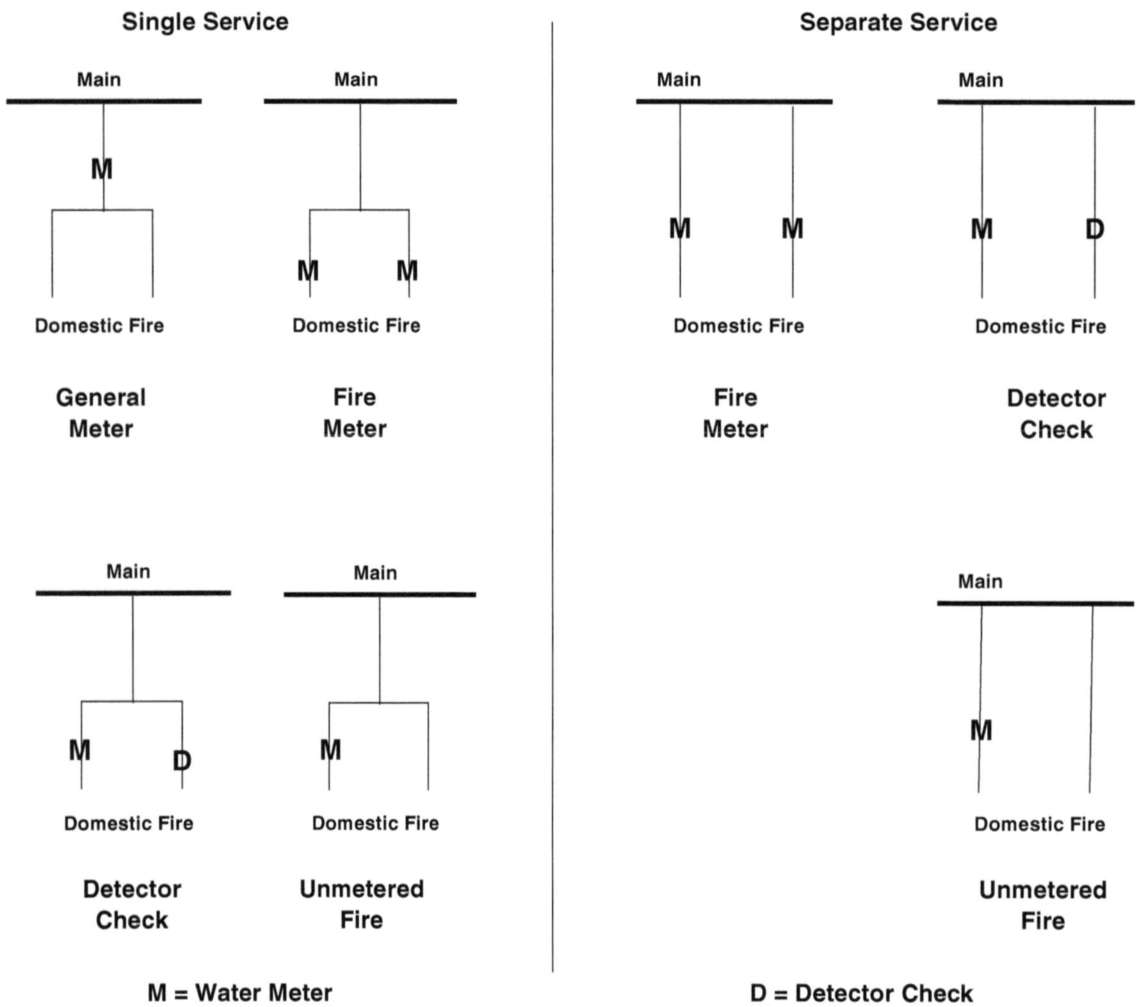

Figure 4-2 Some possible meter arrangements for residential fire service. Per National Fire Protection Association code 13D, the preferable arrangement would be bottom row, second from the left.

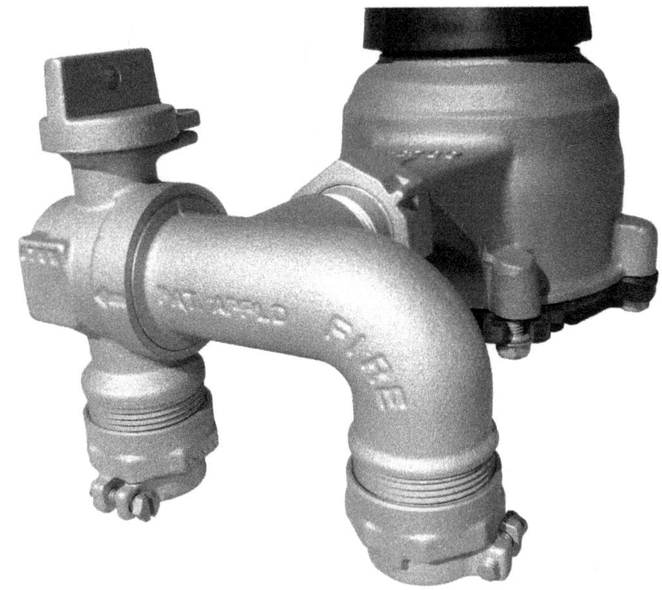

Source: MARS Company

Figure 4-3 Residential fire-service fitting

The selection of a fire-service meter type should follow the general guidelines for type selection for non-fire line meters. Some older commercial/industrial fire line meter applications require a multiregister meter, but a separate fire line makes it possible to get by with a single fire-type commercial meter on the fire line. Figures 4-2 and 4-3 illustrate meter arrangements for residential fire service.

Indoor Settings

The installation of a small meter in a basement or utility room is a relatively simple job. Improper and unsatisfactory settings do occur, however, because of the absence of applicable standards. Whether indoor meters are to be read manually or through some AMR system, a drawing should be made showing basic installation requirements. This drawing should include minimum and maximum elevations above the floor, the type of recommended connections and specification requirements, required valving, and minimum access space required for visual register reading and metering system and setting service work. The drawing should specify that the meter is to be located in the supply line as near as practicable to the point where the water enters the basement or building. If a holding device is not used, the drawing should require an electrical ground connection across the meter. All settings should comply with local codes.

This standard drawing should be furnished to applicants for service, and compliance with the drawing should be one of the conditions for the utility's agreement to serve the premises. This type of drawing provides assurance of uniformly proper installation for the mutual benefit of the utility and the customer.

Outdoor Settings

Many factors influence the design, materials, installation details, and overall performance of an outdoor water-meter installation. Considerations include soil conditions, groundwater level, maximum frost penetration, and accessibility for ease of reading and service. If a pit is used to house the meter, the meter itself should be located at a depth from the surface that

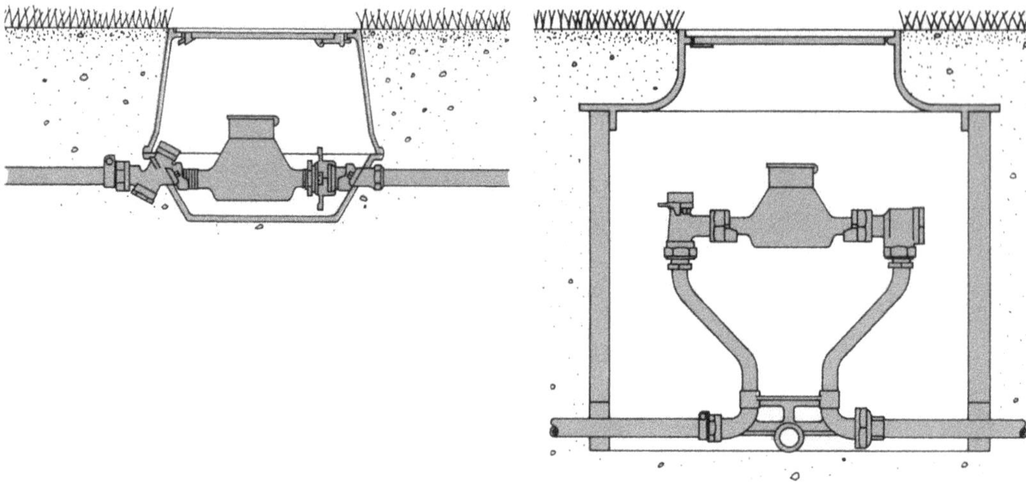

Figure 4-4 Outdoor meter setting with integral yoke (left) and meter yoke (right)

makes reading and accessibility convenient. It is also important to provide 2–4 in. (50–100 mm) of clearance between the service piping and the bottom of the meter pit to avoid any damage or strain that may occur if the meter box settles after installation. Consideration should also be given to the location of the curb stop or service control valve. This valve may be made an integral part of the meter setting or may be located elsewhere and housed in a separate curb box. Figure 4-4 illustrates good outdoor meter-setting practice.

The wide variation in ground frost penetration throughout North America makes it impossible to detail a universally practical outdoor setting. In areas where frost penetration is more than a few inches, serious consideration for frost protection is required. Knowledge of local conditions is necessary to select a pit of sufficient size and depth to provide frost protection. Differences in rate of frost penetration, depending on soil conditions, add considerable complications to this problem. Experience with outdoor installations under a given set of conditions is the best guide for avoiding freezing problems. Suggestions for meter pit design, including recommended size and depth, can be obtained from meter-box manufacturers.

Once sufficient experience has been established for meter-setting standards, it is recommended that a drawing be prepared as a guide for further meter settings. This drawing should specify

- that the meter pit be located as close as possible to the utility line, which is the point of customer delivery;
- that the lid of the pit or box be placed flush with the ground surface;
- that no portion of the riser piping or meter be less than 1–2 in. (25–50 mm) from any portion of the meter box (more if required for frost protection);
- the distance below ground surface where the meter spuds or couplings are to be located;
- the dimensions of the meter box to be used for each size meter; and
- the location of the curb stop or service-control valve.

Outdoor meter setting specifications should be as consistent as possible over an extended period of time to avoid costly modifications or retrofit expenses. Pit/vault lid types and materials decisions are also an important consideration for reading automation and other monitoring devices that might be installed in the meter setting.

Large Meter Settings

By convention, small meters are those 1 in. and smaller, intermediate meters are 1½ in. and 2 in, and large meters are those over 2 in. in line size. Large meter settings, although made less frequently than small meter settings, are relatively expensive and require considerable preliminary planning. Large meters are heavy, and removal for service or testing can be costly and time-consuming. Provisions for fire service must also be given serious consideration.

Some utilities have adopted the practice of installing meters in manifolds of two or more equal branches with meters one size smaller than the main line. These installations may consist of compound, turbine, multijet, positive displacement, singlejet, electromagnetic, or ultrasonic meters. If manifolded, caution should be exercised to ensure both branches contain meters of the same size, type, model, and manufacturer so that the water flow is balanced and performance life of the meters is equal. A manifold installation provides assurance of continuous service because a metered, alternative water flow is available during maintenance or emergency situations. In addition to following the manufacturer's recommendations, Figure 4-5 shows examples of various manifold installations using good installation practice and fire-service provisions.

If a manifold contains two or more branches or if there are meters of different types or sizes, the installation requires considerable care in its design. Flow regulating valves (not shutoff valves) are required for pressure loss adjustments to ensure proper water distribution through the various branches. The meter manufacturers or other engineering experts on hydraulics must be consulted because of the technical issues involved.

Check valves may be necessary in some manifolds to prevent recirculation or to improve the low flow registration accuracy of the manifold system. Use of check valves requires careful consideration and should be designed by a professional.

To determine minimum distance to install subject fittings from velocity type meters such as Class I and/or II turbine meters, singlejet, compound meters, electromagnetic, ultrasonic, and fire-service meters, multiply the nominal pipe diameter of the installation by the appropriate number found in Table 4-1.

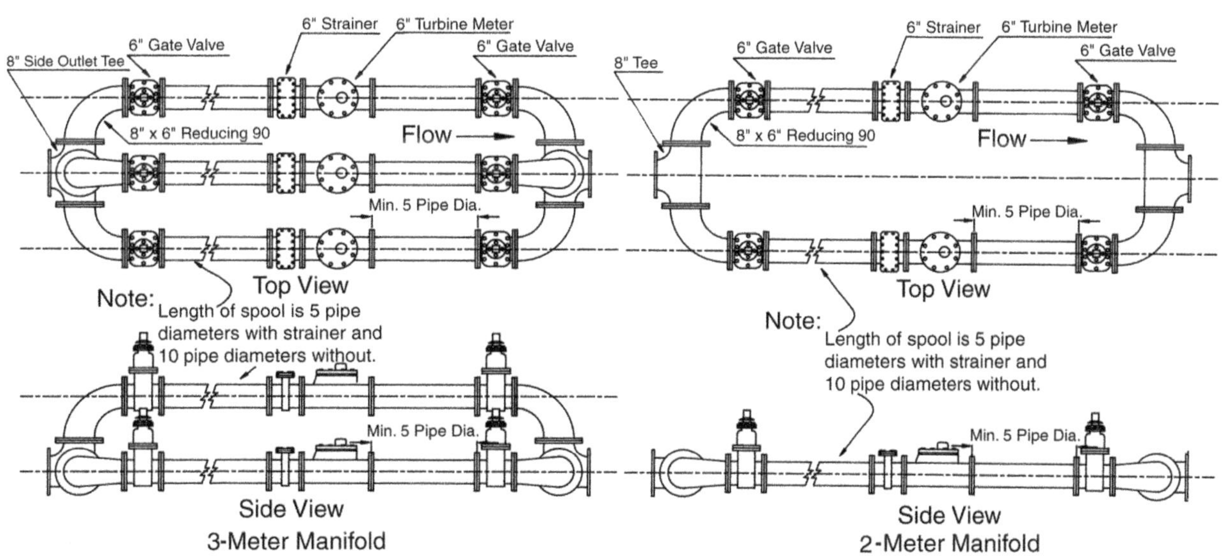

Figure 4-5 Typical 3- and 2-meter hydraulically balanced manifolds

Table 4-1 Large meter installation guides for velocity meters**

Types of Fittings	Distance Upstream in Pipe Diameters	Distance Downstream in Pipe Diameters
Tees and crosses	10	5
Elbows and reducers	10	5
Tees and crosses with strainer or straighteners upstream	5	5
Elbows and reducers with strainer or straighteners upstream	5	5
Angle strainer	5	3
Basket strainer	5	3
Gate valve	1 to 3	1 to 3
Butterfly valve	5	5
Plug valve	5	5
Check valve	*	5
Pressure regulator	*	5
Test tee and plug	*	3
Saddle	*	3

* These fittings are not recommended for use upstream of a water meter. Under extreme high pressures, when it is necessary to protect the meter, pressure regulators may be considered for upstream use only after consulting with the meter manufacturer.

** The manufacturer should be consulted for straight-pipe recommendations.

Custom-built meter setting devices in 1½-in. (40-mm) and 2-in. (50-mm) sizes with built-in bypass systems and valves with locking arrangements are now available. These devices offer a uniformity of meter settings, facilitate meter maintenance, and save space in a typical installation.

Many utilities prefer single-unit installations for large meters and find them very satisfactory. When a single large meter is installed, a bypass circuit should be provided so that meter maintenance can be accomplished without service interruptions. When large meters are installed in a vault, provision should be made for at least 20 in. (500 mm) of clearance to the vertical vault walls and at least 24 in. (600 mm) of head space from the highest point on the meter to the vault cover. Also, when a large meter installation is planned, it is essential that practical testing requirements be carefully considered in the meter and vault layout. Test valves should be installed to permit volumetric field tests, and provisions should be made for discharging test water.

The size and technology type, as well as the meter brand, may require a variety of sizes of test ports for field testing. If a test port provided on the piping or water meter is not adequate to produce the desired high test flow, valved downstream test tees should be incorporated into the installation.

Satisfactory large meter settings can be designed in a variety of ways, depending on specific requirements, code specifications, and individual preferences. When manifolds are used, it is important to make sure each line is hydraulically the same in the manifold. However, it should be noted that these settings represent sizeable investments that will provide long and satisfactory service if adequate planning is done in advance. Valuable advice and installation recommendations can be obtained by contacting the meter manufacturers; this advice should be sought in the preliminary planning stages of a new or unusual installation.

Class I or Class II Turbine-Meter, Compound, Singlejet, Electromagnetic, Ultrasonic, and Fire-Service Installations

Certain precautions are recommended when installing velocity-type meters like turbine or singlejet meters because they work most efficiently when there is a swirl-free, pulsation-free, uniform flow–velocity profile in the pipe immediately upstream of the meter.

For Class I and II turbine, singlejet, electromagnetic, ultrasonic, compound, and fire-service meters, the meter manufacturer should be consulted for the proper orientation of the meter (i.e., whether the meter can be positioned horizontally, vertically, or either). Piping should be arranged to ensure the meter remains full of water at all times and under a positive head. For turbine, singlejet, compound and conventional fire-service meters, elbows, reducers, tees, and crosses installed without a strainer upstream from the meter should be no closer than 10 pipe diameters of straight pipe of the same nominal diameter on the meter upstream and five diameters downstream. As a result of their increased flow capacity, smaller turbine meters, singlejet, electromagnetic, and ultrasonic meters are often used to replace larger-sized meters of other types. When pipe reducers are required in such cases, it is important that only gradual or tapered concentric reducers be used. Care should be exercised that the piping flange gaskets are centered and not protruding into the main flow stream.

Avoid installing check valves or pressure-regulating devices upstream of any large meter. When check valves or pressure-regulating devices are required in the piping system, they should be installed downstream of the meter, preferably at a minimum of five pipe diameters. When backflow prevention devices are required, they should also be installed downstream at a minimum of five pipe diameters. Full-opening ball or gate valves are preferred for the meter set's isolation valves. Butterfly or plug valves may also be used as isolation valves if they are located at a minimum of five pipe diameters upstream and downstream of the meter. If the piping system requires a throttling valve, it should always be downstream of the meter at a minimum of five pipe diameters. It is recommended that a bypass arrangement (either permanent or temporary) be provided to permit uninterrupted customer service during periodic testing or routine maintenance in single-meter-set meter installations. Flushing the lines any time a new installation has occurred or when maintenance has been performed on the line is recommended. To facilitate periodic testing, provision should be made for a test tee or plug at a minimum of three pipe diameters downstream of the meter.

A rigid flat-plate, Z-plate, or V-shaped strainer is recommended to condition the flow stream for proper measurement and to protect any turbine metering element from debris carried in the flow stream. The effective open area of the strainer element should be at least twice the open pipe area of the meter inlet. When angle or regular basket strainers are needed, they should be installed downstream or at a minimum of five pipe diameters upstream. When the piping system is also used for fire service, only a FM-approved or UL Standards & Engagement–listed meter should be used. Some meter types require approved fire-service strainers as part of the fire-service metering package. Fire service–rated strainers should be installed upstream of meter types requiring them. Some newer electromagnetic or ultrasonic fire-service meters are listed as not needing strainers; in such installations, fire-service strainers can be installed downstream of these meters to protect other downstream equipment.

When the piping installation, by necessity, creates a flow swirl in the upstream piping, a flow straightener might be required. At least two types are available: one type incorporates a concentric tube bundle, and the other type uses a system of vanes. Either type of straightener can be installed integrally in the meter or immediately upstream. If a flow straightener is not used, some meter types require the run of straight pipe

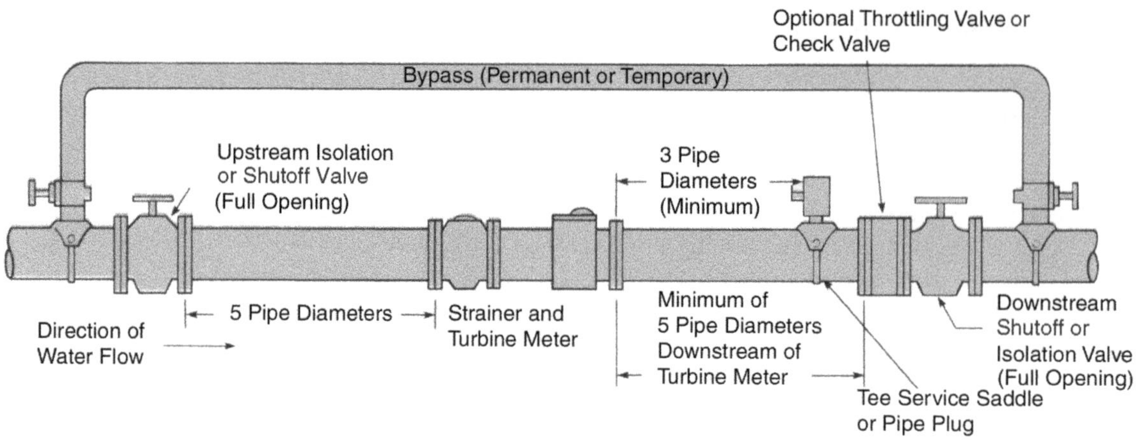

Figure 4-6 Optimum turbine meter installation

immediately upstream of the meter to be increased at a minimum of 10 pipe diameters. The manufacturer's guidelines should be consulted.

Caution should be exercised to avoid entrained air in the meter piping. This is most critical during meter startup or slow, careful recharging of a line following shutoffs necessitated by vacancies or seasonal irrigation meters when large slugs of entrained air could cause damage to the meter's internal mechanical measuring mechanism. Slowly filling the meter piping with a small bleed valve is good practice with the upstream isolation valve open and the downstream isolation valve closed. If possible, the small bleed valve should be located at a high point in the surrounding meter piping. The test opening, if valved, is used for this function.

The installation guidelines described are considered good practices for any meter installation and are repeated again in this section because of their importance in turbine meter installations. Figure 4-6 illustrates many of the suggested installation criteria as applied to Class I and Class II turbine meters.

Electrical Grounded Pipe Systems

The AWWA Policy Statement on Grounding of Electrical Circuits on Water Pipe states,

> AWWA opposes the systematic interchange of stray electric current between electrical distribution systems and water distribution systems, as well as the use of the water pipe system as an essential part of any electrical system. AWWA does not object, however, to the connection of all interior piping of a building to the electrical service neutral and to a separate grounding electrode, provided that such interior pipe systems and grounding connections are electrically isolated from the water utility pipe system. (AWWA 2026)

At the time this edition of Manual M6 was published, the AWWA policy statement on the grounding of electrical circuits to water pipes had last been revised on Jan. 19, 2014, and reaffirmed on Jan. 14, 2017. However, it must be recognized that many pipe systems continue to be used as a grounding electrode system.

In the same vein, Sec. 250.53(D) of the National Electrical Code® requires that continuity of the grounding path or the bonding connection to the interior piping does not rely on water meters. Most utilities require a permanent conductive meter setting or a grounding strapping around meters to prevent accidents to workers changing meters. Meters, both metal and plastic, should be installed in a permanent metallic meter setter or provided with permanently ground-strapped strapping. This is especially important in the case of plastic main cases or plastic meter couplings that are nonconductors of electricity.

REFERENCES

AWWA. 2026. AWWA Policy Statement on Grounding of Electrical Circuits on Water Pipe www.awwa.org/policy-statement/grounding-of-electrical-circuits-on-water-pipe/ (accessed February 5, 2026).

National Fire Protection Association. 2025. *NFPA 70, National Electrical Code*. Quincy, MA: NFPA.

This page intentionally blank.

Chapter 5

Testing of Meters: Test Procedures and Equipment

INTRODUCTION

The only way to determine whether a specific meter is providing acceptable accuracies is to test it. Testing may be conducted for new meters before they are placed into service, as part of the meter repair process, or to verify performance after some amount of time in service. Meter over-registration will result in overbilling of customers, and meter under-registration will result in billing inequities, as well as revenue loss and poor water accountability, requiring a larger proportion of rates to recover nonrevenue water losses. Meter inaccuracies have further impacts when water meter registration is also used to determine charges for other utility services, such as wastewater. Concerns over billing inequities may lead regulatory bodies to establish requirements for meter performance and for the frequency of meter testing. Concerns over billing inequalities, water accountability, and water conservation have led to more stringent regulations for meter testing frequency to improve utilities' ability to identify and apply meter accuracy performance information to these areas of concern.

Mechanical water meters can be subject to wear and deterioration and lose their peak efficiency over a period of time. How long mechanical water meters retain their overall accuracy depends on many factors, such as the quality of the water being measured, rates of flow, total quantity, and chemical buildup and abrasive materials carried by the water. Meters with no moving parts, while not subject to wear, might also experience changes in performance over time. By using reading frequency data from AMR/AMI technology, meter analytics that tracks both the performance of a given meter over time, as well as the performance of its similar or like meters in similar applications and demand profiles, may be able to flag potential changes in meter performance for additional investigation.

The economic advantage of meter maintenance programs has been recorded in many articles. The question remains as to how often meters should be tested. The AWWA meter standards provide testing frequency guidelines as a starting point. An economic analysis depends on factors such as rates charged for water; the effects of differing water qualities on meters; and the cost of purchasing, removing, testing, repairing, and installing meters. A proper economic balance should be attained between the cost of testing and the costs of measurement inaccuracies. If meter accuracies are not adequately maintained, the utility typically loses revenue. Conversely, if the cost of a meter maintenance program (including testing) is more than the loss of revenue incurred, the overall result may be economic waste. Analytics can assist in such decision-making.

Beyond the economics of testing, the water utility will have to be able to provide a way to justify the overall percentage meter error that gets entered into the AWWA Free Water Audit Software under apparent loss, for those utilities that use the audit software. The consequences of misstating this percentage error are that the performance indicators for the audit will be incorrect. In turn, this may cause a utility to work on the wrong water loss remediation, thus wasting valuable resources for loss recovery that may not occur.

ACCURACY LIMITS

Cold water meters for revenue applications have an inherent variation in registration over their entire range of flows. These accuracy percentages can worsen as the meter is subject to conditions noted earlier. As an example, a new mechanical ⅝-in. (15-mm) water meter in good condition will register within the following limits: at least 95 percent at ¼–gal per min (gpm) (0.06-m^3/h) flow, increasing in accuracy to maximum registration of 98.5–101.5 percent near 2 gpm (0.45 m^3/h) (i.e., at roughly 10 percent of rated meter capacity) and then decreasing gradually to not less than 98.5-percent accuracy at the safe maximum rated capacity of 20 gpm (4.5 m^3/h). (Refer to Figure 5-1 for an illustration of a typical accuracy curve for a mechanical meter). Meter manufacturer warranty documents should be reviewed regarding the meter's accuracy performance over its expected life and conditions affecting long-term accuracy.

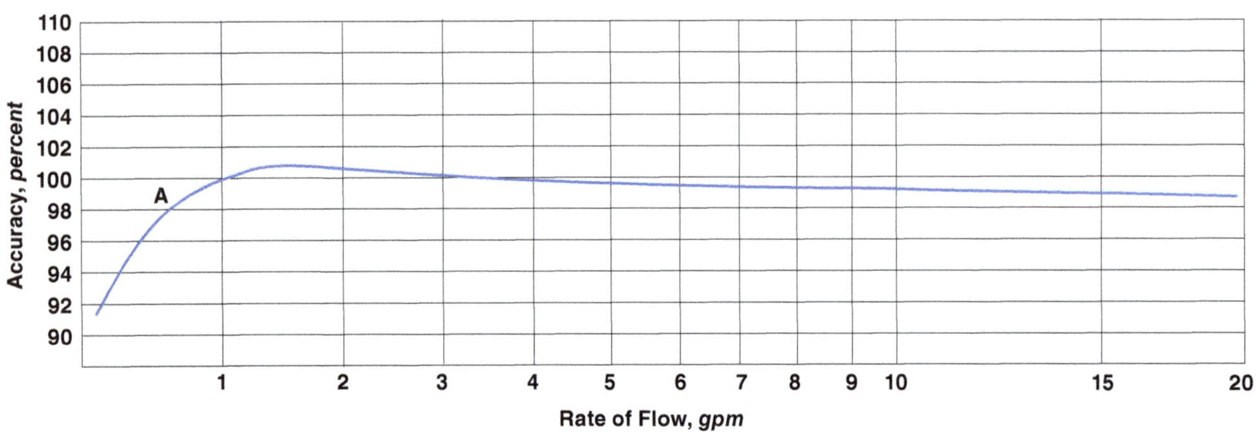

Highest registration occurs at approximately 7 to 10 percent of maximum capacity.

Figure 5-1 A typical accuracy curve for a ⅝-in (15-mm) meter

It may not be economically feasible to repair older mechanical meters to meet the accuracy requirements for new meters. The water utility should carefully weigh all costs to make a specific determination. Smaller sizes of electronic meters cannot typically be repaired. Some models of larger size meters might have replaceable batteries or measurement assemblies that can extend the meter's life. Consult the meter manufacturer's technical documents for such information. Note that separate, reduced accuracy limits will be shown in Table 5-3 for repaired meters on the minimum flow test for some mechanical meter types. These provisions require that repaired meters do not register less than 90 percent on this test. A higher accuracy percentage may be a requirement of the federal, state, provincial, territorial, or local regulatory agencies.

In many cases, meter manufacturers have provided revenue-maintenance programs in the form of replacement measuring-chamber assemblies and modern register assemblies, wherein the accuracy of the rebuilt meter is the same level as that of a new meter. The reality is that reduced accuracies for repaired meters are not acceptable to most modern utility meter shops due to the importance of revenue protection, billing equity, and water accountability. The practice of repairing meters to new meter/rebuilt standards is becoming standard practice.

Accuracy Performance Based on Application

A weighted average meter accuracy can be calculated, based on accuracy test results at various flow rates coupled with a model of actual customer consumption pattern information. These weighting functions can provide an approximation of a meter's actual water accountability in the field. However, accuracies determined at a few test points might not be representative of a specific meter's performance at other flow rates with significant consumption, so additional test points could be needed to improve results. Also, consumption patterns can be highly variable from site to site, particularly in nonresidential applications, so datalogging may be needed to improve the reliability of the weighting function. Consumption patterns can also change over time (due to water conservation measures, repurposing of a site, and other factors) and therefore require periodic reverification.

One such weighting function that has been used for residential meter applications is the algebraic sum of 15 percent of the low flow results, 70 percent of the intermediate flow results, and 15 percent of the maximum flow results. For typical turbine-meter applications, a different weighting function such as 10 percent of the low flow results, 65 percent of the intermediate flow results, and 25 percent of the maximum flow results might be used. For compound and fire-service compound-type meters, the weighted average meter accuracy might be one-third the accuracy results at the maximum test-flow rate of the mainline meter, one-third of the maximum bypass test flow rate, and one-third the intermediate test-flow rates of the bypass meter.

Accuracy Changes Versus Removal From Service

Meters with unacceptable accuracies (as defined by regulatory agencies or by internal concerns over revenue, billing equity, or water accountability) should be repaired or replaced. Table 5-1 provides an example of recommended accuracy limits. (These limits may be driven, in part, by the test uncertainties associated with the methods and equipment being used in determining meter accuracies.) The examples of accuracies in Table 5-1 are not intended to provide a contradiction to guidance from regulatory bodies; however, they may provide guidance for a utility to determine potential repair or replacement.

Table 5-1 Examples of accuracy limits for compliance with guidelines

Meter Type (all sizes)	Accuracy Limits as Found by Testing, *percentage*	
	Normal Test Flow Rates	Minimum Test Flow Rates
Displacement	96–102	90–102
Multijet	96–102	90–104
Singlejet	96–102	90–104
Fluidic-oscillator	96–102	90–102
Propeller and turbine	96–103	Not applicable
Compound and fire-service	95–104	Not applicable
Ultrasonic and electromagnetic	97–103	95–105

Determining the optimum number of years a meter should remain in service between tests is difficult and may need to include considerations such as level of revenue-generated (both water and wastewater) meter size/type/brand, cumulative years in service, consumption patterns, and water quality. Table 5-2 contains testing intervals for many states, while the appendix of most AWWA meter standards provides suggested testing intervals. More frequent test intervals may detect meter inaccuracies in a more timely manner.

Statistical sample testing in a metered distribution system is a method for determining the optimum number of years a specific, homogeneous population of meters should remain in service, especially in residential applications for meters ⅝ (15 mm) through 1 in. (25 mm) in size. Sample testing is a cost-effective management approach to determine the variables affecting meter performance and to monitor the overall accuracy of the metering system. Using established statistical methods, a random selection of meters determined by the year the meter was installed will provide data on the entire metering system. Weighting the statistical sample information with the system flow demand and consumption pattern information determines the service-life decision.

The sample-test database, sorted by the purchase year and by the consumption rate, can be used to identify or window groups of meters with similar accuracy problems. For example, a utility may decide that a given population of meters will remain in service if 95 percent of the sample group complies with the recommended accuracy limits. The actual percentage used might depend upon the return on investment for retaining or replacing the sampled set of meters. Those sampled meters that fall outside of the recommended accuracy limits could also be used as parameters for additional sampling to determine if a more targeted group can be identified for replacement, which might extend the life of the original group after the noncompliant meters are removed from the population. This information will be used as criteria to select replacement schedules based on performance rather than age, type, or consumption. Maintenance programs are an excellent opportunity to upgrade meters with new technology. This ensures that the money used to exchange meters is spent on the poorest performing meters within the system.

As an alternative or supplement to statistical sample testing, data analytics based upon meter reading history can be used to identify meter populations or individual meters that should be tested for possible accuracy degradation.

Table 5-2 State public service commission regulations for periodic testing of water meters as of 8 July 2015*

			Meter Size *in. (mm)*												
			⅝	¾	1	1½	2	3	4	6 or larger	⅝	¾	1	1½	2
			(15)	(20)	(25)	(40)	(50)	(80)	(100)	(150+)	(15)	(20)	(25)	(40)	(50)
State	Effective Year	Rule Number	Interval Between Tests *Years*								Registration Between Tests *1,000 ft^3 (28^{met})*				
Alabama	1988	W-17	10	8	6	4	4	3	2	1					
Alaska (1)															
Arizona (2)	1982	R14-2-408. E													
Arkansas (3)	(4)	7.04	10	8	6	4	4	4	4	4					
California	2009	103-A 3.6	20	20	15	10	10	10	10	10					
Colorado	2010	723-5:5304	10	8	6	4	4	4	4	4					
Connecticut	1978	16-11-88	8	8	8	4	4	3	2	1					
Delaware (5)	2012	38	15	15	10	10	3	3	3	1					
Florida	1986	25-30.265	10	8	6	4	4	3	2	1					
Georgia (6)															
Hawaii (1)															
Idaho (1)															
Illinois	1975	I.A.C Title 83 Part 600.340	10	6	6	4	4	4	4	4	100	300	300		
Indiana	2013	170 IAC 6-1-10	10	8	6	4	4	4	4	4	100	150	300		
Iowa (2)	2014	r.199-21.6(1)													
Kansas (5)	2011	IX.8													
Kentucky	1992	807 KAR 5:066	10	10	10	4	4	2	1	1					
Louisiana (1)															
Maine	1996	35 M.R.S.A 3.3.G	8	8	8	6	6	4	2	1	100	150	300	450	750
Maryland	1968	20.70.06.09	10	8	6	4	4	2	2	1					
Massachusetts (5)	2004			15	15	4	4	2	1	1					
Michigan (6)	2008														
Minnesota (1)															
Mississippi	2012	67.115	10	8	6	4	4	4	4	4	20	300	400		
Missouri	2003	4 CSR 240 10.030(38)	10	8	6	4	4	4	4	4	200	300	400		
Montana (7)	1982	38.5.2513													
Nebraska (1)															
Nevada (1)															
New Hampshire	2013	605.04(c)	10	10	4	4	4	2	1	1					
New Jersey	(4)	14:9-3.2	10	8	6	4	4	4	4	4	750 (8)	1,000 (8)	2,000 (8)		
New Mexico (1)															

(continued)

Table 5-2 State public service commission regulations for periodic testing of water meters as of 8 July 2015* (*Continued*)

			Meter Size *in. (mm)*												
			⅝ (15)	¾ (20)	1 (25)	1½ (40)	2 (50)	3 (80)	4 (100)	6 or larger (150+)	⅝ (15)	¾ (20)	1 (25)	1½ (40)	2 (50)
State	Effective Year	Rule Number	Interval Between Tests *Years*								Registration Between Tests *1,000 ft³ (28met)*				
New York	1981	16 NYCRR Part 500.1	15	15	5	4	4	3	2	1	1500 (8)	2000 (8)	2000 (8)	3000 (8)	7,000 (8)
North Carolina (9)	1952	R7-32				4	4	3	2	1					
North Dakota (1)															
Ohio (3)	2008	OAC 4901:1-15-19	3	3	3	3	3	3	3	3					
Oklahoma	1999	165:65-7-11	10	10	10	6	6	4	2	2					
Oregon (2)	(4)	OAR 860-036-0110													
Pennsylvania (3)	1998	65.8(b)	12	8	6	4	4	4	4	4	100	150	300		
Rhode Island	1966	Schedule	10	10	10	10	10	2	1	1					
South Carolina (1)															
South Dakota (1)															
Tennessee (2)	1974	1220-4-3-34													
Texas (1)															
Utah (2)	2015	R746-330-3(B)													
Vermont	1973	CVR 30.3.24.c	5	5	5	5	5	5	5	5					
Virginia (1)	1972	§ 56-245.1													
Washington (1)	1999	480-110-405													
West Virginia	2011	150CSR 7-6.4.a	10	10	7	5	5	3	1	1					
Wisconsin	2001	PSC 185.76	10	10	10	4	4	2	2	1					
Wyoming	1979	Ch. 6 Sec. 608	10	8	6	4	4	4	4	4	100	150	300		

* Data are condensed for simplification. Detailed or specific information should be obtained from the state regulatory agency. Testing intervals included in this tabulation are statutory and are not recommended periods.
This table may not represent electronic meter technologies. The utility can consult their state for any updates to the regulations.
(1) No regulations for periodic testing having been adopted.
(2) Each utility sets its own schedule.
(3) Testing is free upon customer request at these time intervals.
(4) Effective year not furnished.
(5) Data collected from independent utility. May not be statewide commission.
(6) Individual utilities adopting AWWA standards.
(7) Utility will pull 5 percent of all meters annually for registration accuracy testing.
(8) Registration is in units of 1,000 gal.
(9) Applies only to compound, velocity, and fire-line meters.

Change-Gear Calibration

Decades ago, numerous mechanical meter designs included registration gear sets that could be modified in the field. This would allow recalibration, moving the entire meter accuracy curve up or down by changing the cumulative reduction ratio of these gear sets. Such designs are no longer relevant.

Multijet Meters: Calibration Device

In some multijet meters, changes are made hydraulically. Calibration is done by a variable-port regulating device incorporated along an internal bypass channel in some meter designs. Usually the device will be in the form of a screw, accessible from the outside without the need to disassemble the meter or the register and protected by a sealed plug.

Turbine Meters: Calibration Device

Some turbine meter designs have an adjusting vane that can be used for a simple hydraulic calibration. Adjustment is done with a screwdriver. Changing the position of the adjusting vane will affect the entire accuracy curve, but it might not shift the whole curve by the same amount.

Electronic Meters: Calibration

In many cases, electronic meters are calibrated by the manufacturer. Utilities should consult with specific meter manufacturers on any calibration requirements or options available.

Postcalibration

After any calibration activity, it is recommended that postadjustment meter accuracy testing be performed to verify meter accuracy is within acceptable limits.

TESTING METERS IN A TEST FACILITY

New Meters

New meters should be tested for accuracy of registration at flow rates and test-flow quantities in accordance with Table 5-3 before they are placed in service (also known as acceptance testing). A utility may choose to test only a sample of new meters since it may not be cost-effective or feasible to test all meters. (A utility may consider additional test flow rates based upon their own specifications or the claims of the meter manufacturer.) It is advisable to test new meters in a timely manner. This procedure ensures the water utility that the new meter meets all accuracy requirements and that a baseline history is available when the meter is eventually retested or if a customer challenges the meter's accuracy. Through the analytics of well-constructed meter accuracy change data, projections on changes in billing amounts following repairs or replacement can be modeled.

During the procurement process, specifications that require the manufacturer to provide certified meter test results are advantageous. Certified test results may easily be transferred to a database, as a step in establishing the complete history of a meter.

The utility can conduct its own acceptance testing or contract with a third-party testing service. Given inherent test uncertainties, the utility should expect some variation in results between different test facilities or between multiple tests at a single facility (see discussion on test uncertainties later in this chapter). Sampling on a statistical basis can be an efficient alternative to testing every single meter in a new meter shipment.

Meters Removed From Service

After some time in service, meters may be tested as-is or after repair/rebuild activities. Conducting an accuracy test prior to repairs or removal from service is a good practice because it provides the utility with a basis for addressing customer billing issues (i.e., because the removed meter most likely was replaced with a higher-accuracy meter). Testing of such meters in a dedicated test facility can be more efficient and typically results in smaller test uncertainties when compared to in-situ field testing. This can be particularly true if meter repairs are also attempted. However, field testing does have the advantage of incorporating installation-dependent effects upon meter accuracy and, in avoiding removal of the meter from its field setting, does not disturb the internal surfaces of the meter or possibly introduce debris into the meter. Any type of meter that has more exacting upstream and downstream piping requirements or is extremely expensive to remove from service might be a candidate for field testing.

In testing a meter that has been removed from service, care must be taken to avoid introducing particulates and other contaminants into a test facility. Conversely, the interior of the meter should be maintained at (or restored to) conditions similar to those that were present in the field, for valid as-is testing. Good practice is to maintain residual water in the meter as it is being pulled and promptly installing end caps to prevent the ingress of debris while keeping the chamber lubricated. As a general rule, meters removed from service should be treated like a new meter to avoid introduction of uncertainty into the testing process.

Program Coordination

The scope of a meter test program must be determined and assessed against the resources of both test facility staffing and equipment capacity. Equipment for testing of residential meters, intermediate meters (e.g., 1½–2 in.), and larger meters may have independent scheduling constraints. The utility must establish its own priorities or needs with respect to testing of new meters and for testing of meters pulled from field service. For the latter, regulatory requirements, recommended time between tests from AWWA standards or from prior utility-specific experience, or a statistical sampling plan to assess field behavior may drive the number of meters to be tested. A utility might also prioritize testing of meters used at sites with the highest revenues.

When testing meters pulled from the field, there must be coordination with field service staff who remove the meters, as well as the stocking of replacement meters for these sites. For utilities that do so, staff, material, and equipment must be arranged for meter repairs, and subsequent retesting of these meters must be coordinated.

Table 5-3 Test requirements for new, rebuilt, and repaired cold-water meters*

Displacement Meters (ANSI/AWWA C700 and C710)

	Maximum Rate (All Meters)				Intermediate Rate (All Meters)				Minimum Rate (New and Rebuilt)				Minimum (Repaired)
Size, *in.*	Flow Rate,[†] *gpm*	Test Quantity[††] *gal*	Test Quantity[††] *ft*^{3}	Accuracy Limits, *percentage*	Flow Rate,** *gpm*	Test Quantity[††] *gal*	Test Quantity[††] *ft*^{3}	Accuracy Limits, *percentage*	Flow Rate, *gpm*	Test Quantity[††] *gal*	Test Quantity[††] *ft*^{3}	Accuracy Limits, *percentage*	Accuracy Limits, *percentage (min)*
½	8	100	10	98.5–101.5	2	10	1	98.5–101.5	¼	10	1	95–101.5	90
½ × ¾	8	100	10	98.5–101.5	2	10	1	98.5–101.5	¼	10	1	95–101.5	90
⅝	15	100	10	98.5–101.5	2	10	1	98.5–101.5	¼	10	1	95–101.5	90
⅝ × ¾	15	100	10	98.5–101.5	2	10	1	98.5–101.5	¼	10	1	95–101.5	90
¾	25	100	10	98.5–101.5	3	10	1	98.5–101.5	½	10	1	95–101.5	90
1	40	100	10	98.5–101.5	4	10	1	98.5–101.5	¾	10	1	95–101.5	90
1½	50	100	10	98.5–101.5	8	100	10	98.5–101.5	1½	100	10	95–101.5	90
2	100	100	10	98.5–101.5	15	100	10	98.5–101.5	2	100	10	95–101.5	90

Multijet Meters (ANSI/AWWA C708)

	Maximum Rate (All Meters)				Intermediate Rate (All Meters)				Minimum Rate (New and Rebuilt)				Minimum (Repaired)
Size, *in.*	Flow Rate,[†] *gpm*	Test Quantity[††] *gal*	Test Quantity[††] *ft*^{3}	Accuracy Limits, *percentage*	Flow Rate,** *gpm*	Test Quantity[††] *gal*	Test Quantity[††] *ft*^{3}	Accuracy Limits, *percentage*	Flow Rate, *gpm*	Test Quantity[††] *gal*	Test Quantity[††] *ft*^{3}	Accuracy Limits, *percentage*	Accuracy Limits, *percentage (min)*
⅝	15	100	10	98.5–101.5	1	10	1	98.5–101.5	¼	10	1	97–103	90
⅝ × ¾	15	100	10	98.5–101.5	1	10	1	98.5–101.5	¼	10	1	97–103	90
¾	25	100	10	98.5–101.5	2	10	1	98.5–101.5	½	10	1	97–103	90
1	35	100	10	98.5–101.5	3	10	1	98.5–101.5	¾	10	1	97–103	90
1½	70	100	10	98.5–101.5	5	100	10	98.5–101.5	1½	100	10	97–103	90
2	100	100	10	98.5–101.5	8	100	10	98.5–101.5	2	100	10	97–103	90

Singlejet Meters (ANSI/AWWA C712)

	Maximum Rate (All Meters)				Intermediate Rate (All Meters)				Minimum Rate (New and Rebuilt)				Minimum (Repaired)
Size, *in.*	Flow Rate,[†] *gpm*	Test Quantity[††] *gal*	Test Quantity[††] *ft*^{3}	Accuracy Limits, *percentage*	Flow Rate,** *gpm*	Test Quantity[††] *gal*	Test Quantity[††] *ft*^{3}	Accuracy Limits, *percentage*	Flow Rate, *gpm*	Test Quantity[††] *gal*	Test Quantity[††] *ft*^{3}	Accuracy Limits, *percentage*	Accuracy Limits, *percentage (min)*
⅝	15	100	10	98.5–101.5	2	10	1	98.5–101.5	¼	10	1	95–101.5	90
⅝ × ¾	15	100	10	98.5–101.5	2	10	1	98.5–101.5	¼	10	1	95–101.5	90
¾	25	100	10	98.5–101.5	3	10	1	98.5–101.5	½	10	1	95–101.5	90
1	40	100	10	98.5–101.5	4	10	1	98.5–101.5	¾	10	1	95–101.5	90
1½	50	100	10	98.5–101.5	8	100	10	98.5–101.5	½	100	10	95–101.5	90
2	100	100	10	98.5–101.5	15	100	10	98.5–101.5	½	100	10	95–101.5	90
3	160	500	50	98.5–101.5	20	100	10	98.5–101.5	½	100	10	95–101.5	90
4	250	500	50	98.5–101.5	40	100	10	98.5–101.5	¾	100	10	95–101.5	90
6	500	1,000	100	98.5–101.5	60	100	10	98.5–101.5	1½	100	10	95–101.5	90

(continued)

Table 5-3 Test requirements for new, rebuilt, and repaired cold-water meters* (*Continued*)

Fluidic-Oscillator Meters (ANSI/AWWA C713)

	Maximum Rate (All Meters)				Intermediate Rate (All Meters)				Minimum Rate (New and Rebuilt)				Minimum (Repaired)
		Test Quantity††				Test Quantity††				Test Quantity††			Accuracy Limits,
Size, *in.*	Flow Rate,† *gpm*	*gal*	*ft³*	Accuracy Limits, *percentage*	Flow Rate,** *gpm*	*gal*	*ft³*	Accuracy Limits, *percentage*	Flow Rate, *gpm*	*gal*	*ft³*	Accuracy Limits, *percent*	*percentage (min)*
½	8	100	10	98.5–101.5	2	10	1	98.5–101.5	¼	10	1	95–101.5	90
½ × ¾	8	100	10	98.5–101.5	2	10	1	98.5–101.5	¼	10	1	95–101.5	90
⅝	15	100	10	98.5–101.5	2	10	1	98.5–101.5	¼	10	1	95–101.5	90
⅝ × ¾	15	100	10	98.5–101.5	2	10	1	98.5–101.5	¼	10	1	95–101.5	90
¾	25	100	10	98.5–101.5	3	10	1	98.5–101.5	½	10	1	95–101.5	90
1	40	100	10	98.5–101.5	4	10	1	98.5–101.5	¾	10	1	95–101.5	90
1½	50	100	10	98.5–101.5	8	100	10	98.5–101.5	1½	100	10	95–101.5	90
2	100	100	10	98.5–101.5	15	100	10	98.5–101.5	2	100	10	95–101.5	90

Class I Turbine Meters, Vertical-Shaft Type (ANSI/AWWA C701)

	Maximum Rate (All Meters)				Intermediate Rate (All Meters)				Minimum Rate (New and Rebuilt)				Minimum (Repaired)
		Test Quantity††				Test Quantity***				Test Quantity††			Accuracy Limits,
Size, *in.*	Flow Rate,† *gpm*	*gal*	*ft³*	Accuracy Limits, *percentage*	Flow Rate,** *gpm*	*gal*	*ft³*	Accuracy Limits, *percentage*	Flow Rate, *gpm*	*gal*	*ft³*	Accuracy Limits, *percentage*	*percentage (min)*
¾	30	100	10	98–102	3	10	1	98–102	1½	10	1	98–102	—
1	50	100	10	98–102	5	10	1	98–102	2	10	1	98–102	—
1½	100	500	50	98–102	10	100	10	98–102	3	100	10	98–102	—
2	160	500	50	98–102	16	100	10	98–102	4	100	10	98–102	—
3	350	1,000	100	98–102	35	100	10	98–102	6	100	10	98–102	—
4	600	1,500	200	98–102	60	100	10	98–102	8	100	10	98–102	—
6	1,250	4,000	500	98–102	125	1,000	100	98–102	15	1,000	100	98–102	—

Class II Turbine Meters, In-Line (High-Velocity) Type (ANSI/AWWA C701)

	Maximum Rate (All Meters)				Intermediate Rate (All Meters)				Minimum Rate (New and Rebuilt)				Minimum (Repaired)
		Test Quantity††				Test Quantity***				Test Quantity††			Accuracy Limits,
Size, *in.*	Flow Rate,† *gpm*	*gal*	*ft³*	Accuracy Limits, *percentage*	Flow Rate,** *gpm*	*gal*	*ft³*	Accuracy Limits, *percentage*	Flow Rate, *gpm*	*gal*	*ft³*	Accuracy Limits, *percentage*	*percentage, (min)*
1½	100	500	50	98.5–101.5					4	100	10	98.5–101.5	—
2	160	500	50	98.5–101.5					4	100	10	98.5–101.5	—
3	350	1,000	100	98.5–101.5					8	100	10	98.5–101.5	—
4	630	1,500	200	98.5–101.5					15	100	10	98.5–101.5	—
6	1,400	4,000	500	98.5–101.5					30	1,000	100	98.5–101.5	—
8	2,400	7,000	900	98.5–101.5					50	1,000	100	98.5–101.5	—
10	3,800	10,000	1,300	98.5–101.5					75	1,000	100	98.5–101.5	—
12	5,000	15,000	2,000	98.5–101.5					120	1,000	100	98.5–101.5	—
16	7,000	30,000	4,000	98.5–101.5					200	1,000	100	98.5–101.5	—
20	10,000	40,000	5,000	98.5–101.5					300	1,000	100	98.5–101.5	—

(continued)

Table 5-3 Test requirements for new, rebuilt, and repaired cold-water meters* (*Continued*)

Propeller Meters (ANSI/AWWA C704)

	Maximum Rate (All Meters)				Intermediate Rate (All Meters)				Minimum Rate (New and Rebuilt)				Minimum (Repaired)
		Test Quantity††				Test Quantity††***				Test Quantity††			Accuracy Limits,
Size, *in.*	Flow Rate,† *gpm*	*gal*	*ft³*	Accuracy Limits, *percentage*	Flow Rate,** *gpm*	*gal*	*ft³*	Accuracy Limits, *percentage*	Flow Rate, *gpm*	*gal*	*ft³*	Accuracy Limits, *percentage*	*percentage (min)*
2	100	300	40	98–102					45	200	25	98–102	90
3	250	800	100	98–102					80	200	25	98–102	90
4	500	1,500	200	98–102					85	250	30	98–102	90
6	1,200	2,500	300	98–102					160	500	60	98–102	90
8	1,500	3,000	400	98–102					190	500	60	98–102	90
10	2,000	4,000	500	98–102					260	500	60	98–102	90
12	2,800	6,000	800	98–102					275	750	100	98–102	90
14	3,750	8,000	1,000	98–102					350	1,000	130	98–102	90
16	4,750	10,000	1,300	98–102					450	1,500	200	98–102	90
18	5,625	12,000	1,600	98–102					550	2,000	250	98–102	90
20	6,875	15,000	2,000	98–102					650	2,500	300	98–102	90
24	10,000	20,000	2,500	98–102					1,000	4,000	500	98–102	90
30	15,000	30,000	4,000	98–102					1,600	6,000	800	98–102	90
36	20,000	40,000	5,000	98–102					2,400	7,500	1,000	98–102	90
42	28,000	40,000	5,000	98–102					2,800	10,000	1,300	98–102	90
48	35,000	50,000	6,000	98–102					3,500	12,500	1,500	98–102	90
54	45,000	60,000	8,000	98–102					5,000	16,000	2,000	98–102	90
60	60,000	70,000	9,000	98–102					6,000	20,000	2,500	98–102	90
66	75,000	80,000	11,000	98–102					7,500	25,000	3,000	98–102	90
72	90,000	90,000	12,000	98–102					9,000	28,000	3,500	98–102	90

Compound Meters (ANSI/AWWA C702)§

	Maximum Rate (All Meters)					Change Over Point (All Meters)				Minimum Rate (New and Rebuilt)				Minimum (Repaired)
		Test Quantity††					Test Quantity††				Test Quantity††			Accuracy Limits,
Size, *in.*	Flow Rate,† *gpm*	*gal*	*ft³*	Accuracy Limits, *percentage*		Flow Rate,** *gpm*	*gal*	*ft³*	Accuracy Limits, *percentage*	Flow Rate, *gpm*	*gal*	*ft³*	Accuracy Limits,††† *percentage*	*percentage (min)*
				Class I	Class II									
2	160	400	50	97–103	98.5–101.5				90–103					90
3	320	1,000	100	97–103	98.5–101.5				90–103					90
4	500	1,500	200	97–103	98.5–101.5				90–103					90
6	1,000	3,000	400	97–103	98.5–101.5				90–103					90
8	1,600	4,000	500	97–103	98.5–101.5				90–103					90
10	2,300	4,000	500	97–103	98.5–101.5				90–103					90

(continued)

Table 5-3 Test requirements for new, rebuilt, and repaired cold-water meters* (*Continued*)

Fire-Service Type II (ANSI/AWWA C703)
(Test at intermediate rate not necessary.)[§]

	Maximum Rate (All Meters)					Change Over Point (All Meters)				Minimum Rate (New and Rebuilt)				Minimum (Repaired)
Size, *in.*	Flow Rate,[†] *gpm*	Test Quantity[††] *gal*	Test Quantity[††] *ft³*	Accuracy Limits, *percentage* Type I	Accuracy Limits, *percentage* Type II	Flow Rate,** *gpm*	Test Quantity[††] *gal*	Test Quantity[††] *ft³*	Accuracy Limits, *percentage*	Flow Rate, *gpm*	Test Quantity[††] *gal*	Test Quantity[††] *ft³*	Accuracy Limits,[†††] *percentage*	Accuracy Limits *percentage, min*
3	350	700	100	97–103	98.5–101.5									90
4	700	1,500	200	97–103	98.5–101.5				Not less than 85%				Not less than 95%	90
6	1,600	3,000	400	97–103	98.5–101.5									90
8	2,800	5,000	700	97–103	98.5–101.5									90
10	4,400	9,000	1,200	97–103	98.5–101.5									90

Fire-Service Type, Type III (ANSI/AWWA C703)

	Maximum Rate (All Meters)				Intermediate Rate (All Meters)				Minimum Rate (New and Rebuilt)				Minimum (Repaired)
Size, *in.*	Flow Rate,[†] *gpm*	Test Quantity[††] *gal*	Test Quantity[††] *ft³*	Accuracy Limits, *percentage*	Flow Rate,** *gpm*	Test Quantity*** *gal*	Test Quantity*** *ft³*	Accuracy Limits, *percentage*	Flow Rate, *gpm*	Test Quantity[††] *gal*	Test Quantity[††] *ft³*	Accuracy Limits, *percentage*	Accuracy Limits, *percentage (min)*
3	350	700	100	98.5–101.5	10	100	10	98.5–101.5	4	100	10	95–101.5	—
4	700	1,500	200	98.5–101.5	30	500	50	98.5–101.5	10	100	10	95–101.5	—
6	1,600	3,000	400	98.5–101.5	60	1000	100	98.5–101.5	20	1000	100	95–101.5	—
8	2,800	5,000	700	98.5–101.5	70	1000	100	98.5–101.5	30	1000	100	95–101.5	—
10	4,400	9,000	1,200	98.5–101.5	110	1000	100	98.5–101.5	35	1000	100	95–101.5	—

Electromagnetic and Ultrasonic Meters for Revenue Applications, Type I (ANSI/AWWA C715)

	Maximum Rate (All Meters)				Intermediate Rate (All Meters)				Minimum Rate (New and Rebuilt)				Minimum (Repaired)
Size, *in.*	Flow Rate,[†] *gpm*	Test Quantity[††] *gal*	Test Quantity[††] *ft³*	Accuracy Limits, *percentage*	Flow Rate,** *gpm*	Test Quantity[††] *gal*	Test Quantity[††] *ft³*	Accuracy Limits, *percentage*	Flow Rate,[§§] *gpm*	Test Quantity[††] *gal*	Test Quantity[††] *ft³*	Accuracy Limits,[§§] *percentage*	Accuracy Limits, *percentage (min)*
½	8	100	10	98.5–101.5	0.35	10	1	98.5–101.5	0.11 (0.18)	10	1	95–105 (98.5–101.5)	—
⅝	15	100	10	98.5–101.5	0.4	10	1	98.5–101.5	0.13 (0.20)	10	1	95–105 (98.5–101.5)	—
¾	25	100	10	98.5–101.5	1	10	1	98.5–101.5	0.15 (0.5)	z10	1	95–105 (98.5–101.5)	—
1	40	200	20	98.5–101.5	1.5	10	1	98.5–101.5	0.3 (0.75)	10	1	95–105 (98.5–101.5)	—
1½	60	200	50	98.5–101.5	4	100	10	98.5–101.5	0.6 (2)	100	10	95–105 (98.5–101.5)	—
2	100	500	50	98.5–101.5	5	100	10	98.5–101.5	1 (2.5)	100	10	95–105 (98.5–101.5)	—
3	200	1,000	100	98.5–101.5	15	100	10	98.5–101.5	2.5 (7.5)	100	10	95–105 (98.5–101.5)	—
4	400	2,000	200	98.5–101.5	20	500	50	98.5–101.5	3.5 (10)	300	40	95–105 (98.5–101.5)	—

(continued)

Table 5-3 Test requirements for new, rebuilt, and repaired cold-water meters* (*Continued*)

Electromagnetic and Ultrasonic Meters for Revenue Applications, Type I (ANSI/AWWA C715) (*Continued*)

	Maximum Rate (All Meters)				Intermediate Rate (All Meters)				Minimum Rate (New and Rebuilt)				Minimum (Repaired)
Size, *in.*	Flow Rate,† *gpm*	Test Quantity†† *gal*	Test Quantity†† *ft³*	Accuracy Limits, *percentage*	Flow Rate,** *gpm*	Test Quantity†† *gal*	Test Quantity†† *ft³*	Accuracy Limits, *percentage*	Flow Rate,§§ *gpm*	Test Quantity†† *gal*	Test Quantity†† *ft³*	Accuracy Limits,§§ *percentage*	Accuracy Limits, *percentage (min)*
6	800	5,000	500	98.5–101.5	40	1,000	100	98.5–101.5	9 (20)	300	40	95–105 (98.5–101.5)	—
8	1,000	5,000	500	98.5–101.5	80	3,000	400	98.5–101.5	18 (40)	2,000	300	95–105 (98.5–101.5)	—

Electromagnetic and Ultrasonic Meters for Revenue Applications, Type II (ANSI/AWWA C715)

	Maximum Rate (All Meters)				Intermediate Rate (All Meters)				Minimum Rate (New and Rebuilt)				Minimum (Repaired)
Size, *in.*	Flow Rate,† *gpm*	Test Quantity†† *gal*	Test Quantity†† *ft³*	Accuracy Limits, *percentage*	Flow Rate,** *gpm*	Test Quantity†† *gal*	Test Quantity†† *ft³*	Accuracy Limits, *percentage*	Flow Rate, *gpm*	Test Quantity†† *gal*	Test Quantity†† *ft³*	Accuracy Limits, *percentage*	Accuracy Limits, *percentage (min)*
½	8	100	10	98.5–101.5	2	10	1	98.5–101.5	0.25	10	1	95–105	—
⅝	15	100	10	98.5–101.5	2	10	1	98.5–101.5	0.25	10	1	95–105	—
¾	25	100	10	98.5–101.5	4	20	2	98.5–101.5	0.5	10	1	95–105	—
1	40	200	20	98.5–101.5	6	20	5	98.5–101.5	0.75	10	1	95–105	—
1½	100	500	50	98.5–101.5	8	100	10	98.5–101.5	1.5	100	10	95–105	—
2	150	500	100	98.5–101.5	12	100	10	98.5–101.5	2	100	10	95–105	—
3	350	2,000	200	98.5–101.5	30	100	20	98.5–101.5	4	100	10	95–105	—
4	600	2,000	500	98.5–101.5	50	500	50	98.5–101.5	7.5	300	40	95–105	—
6	800	5,000	500	98.5–101.5	110	1,000	100	98.5–101.5	15	300	40	95–105	—
8	2,000	10,000	1,000	98.5–101.5	200	3,000	400	98.5–101.5	30	2,000	300	95–105	—
10	3,000	10,000	2,000	98.5–101.5	300	5,000	500	98.5–101.5	50	3,000	400	95–105	—
12	4,000	20,000	2,000	98.5–101.5	400	5,000	500	98.5–101.5	65	3,000	400	95–105	—
16	5,500	50,000	7,000	98.5–101.5	750	30,000	4,000	98.5–101.5	110	20,000	3,000	95–105	—
20	8,000	50,000	7,000	98.5–101.5	1,100	30,000	4,000	98.5–101.5	175	20,000	3,000	95–105	—

Meters for Residential Fire Sprinkler Systems in One- and Two-Family Dwellings and Manufactured Homes (ANSI/AWWA C714)

	Maximum Rate (All Meters)				Intermediate Rate (All Meters)				Minimum Rate (New and Rebuilt)				Minimum (Repaired)
Size, *in.*	Flow Rate,† *gpm*	Test Quantity†† *gal*	Test Quantity†† *ft³*	Accuracy Limits, *percentage*	Flow Rate,** *gpm*	Test Quantity†† *gal*	Test Quantity†† *ft³*	Accuracy Limits, *percentage*	Flow Rate,§§ *gpm*	Test Quantity†† *gal*	Test Quantity†† *ft³*	Accuracy Limits, *percentage*	Accuracy Limits, *percentage (min)*
¾	25	100	10	98–102	3	10	1	98–102	½	10	1	95–103	90
1	40	100	10	98–102	4	10	1	98–102	¾	10	1	95–103	90
1½	50	100	10	98–102	8	100	10	98–102	1½	100	10	95–103	90
2	100	100	10	98–102	15	100	10	98–102	2	100	10	95–103	90

* A rebuilt meter is one that has had the measuring element replaced with a factory-made new unit. A repaired meter is one that has had the old measuring element cleaned and refurbished in a utility repair shop.

† These are suggested test flows and test quantities. Testing for high rates of flow can be achieved by testing the meter at 25 percent of the meters rating if the manufacturer's original test certificate indicates a linear curve between 25 and 100 percent of the rated flow range.

(continued)

†† Depending on test conditions and specific meter design details, these test quantities may result in large test uncertainties. Quantity should be one or more full revolutions of the test hand/initial dial (see "maximum indication on initial dial" values specified in mechanical meter standards or a quantity not less than the "maximum allowable indication on initial display" values specified in ANSI/AWWA C715. Quantity should result in not less than 3 min running. Smaller test drafts can be used but may result in repeatability values outside the expected accuracy range. Repeatability testing (three runs at the same test conditions) can determine if too-small test quantities are being used, meriting an increase in test quantities. When limited test capabilities force the use of too-small test quantities, the resultant increase in total test uncertainties and errors need to be recognized when establishing acceptance criteria tolerance.

§ The bypass meter should be tested in accordance with the appropriate test requirements for the type of meter used.

** As this rate varies according to manufacturer, it should be determined for each type of meter tested.

§§ Minimum test flow rates with corresponding accuracy limits in parentheses ( ) are alternative values to accommodate concerns over test times or concerns over test benches being able to conduct testing at the smaller flow rates (without parentheses) reliably. Note that the flow rate values in parentheses are the lowest rates for the normal test limits from ANSI/AWWA C715, and as such would have accuracy limits of 98.5–101.5 percent. If these flow rates in parentheses are used in minimum rate testing, then the recommended test flow rates for the intermediate rate should be twice those listed in the table. However, see the preceding footnote, detailing the preferred use of an intermediate rate determined for each specific make/model.

*** As an alternative to the recommended flow rate, the utility can use a flow rate that is appropriate for their use case.

††† Accuracy limits should follow those of the meter bypass type.

NOTE: Metric conversions: in. × 25.4 = mm, gal × 0.003785 = m^3, gpm × 0.2268 = m^3/h, ft^3 × 0.02831 = m^3.

gpm = gallon per minute

Large Meters

It is generally considered advisable to provide for more frequent testing of large meters because an error in their registration affects revenue to a much greater extent. Some form of testing program is essential, with the actual test interval depending on utility-specific conditions such as accuracy changes over time and financial rate of return. For large utilities, a typical time interval between tests is 1 year.

If enough 3-in. (80-mm) and larger meters are installed, the repair and testing of these larger meters may be delegated to an appropriately sized crew for the situation. They will develop special skills that are necessary for the effective maintenance of larger meters. In any meter-testing program, accurate and readily available records are essential. A formal, ongoing meter record program should be established as an initial step in the program. Electronic data processing has proven to be a highly effective tool in maintaining an effective meter record program and consolidating data for analytics.

When any testing program is considered, meters with the longest service time and those with the heaviest volume should be given priority, as they may have the greatest accuracy degradation (and for the latter, have greater revenue impact).

Probably the best advice that can be given regarding a meter testing program is to be alert to and study all phases of the metering field; there is no substitute for experience in determining the best procedure for any one utility. While a metered system is the first step in equitably spreading the cost of water service, serious inequities can still occur unless all meters are maintained at a high, uniform level of efficiency within the limits of what is economically feasible. Analysis of carefully vetted reading, billing, and testing data is a useful tool in identifying potential meter inaccuracies for additional investigation.

Test Procedures

Test practices can vary widely between utilities and may even vary within a single utility. Uniform test practices, informed by the meter performance requirements of AWWA meter standards, can be established by following the test practice recommendations of this manual. The AWWA meter standards themselves provide specific values for the minimum flow rate testing, but additional test flows, along with test quantities (i.e., test volumes or test draft sizes), are recommended based upon realistic expectations for meter behavior, test resources, and costs, as well as acceptable uncertainties in the determination of accuracies. Test practices might be somewhat different for new meters versus meters pulled from the field, but some similarities need to be maintained in order to assess changes in meter performance over time. Test practices may be modified to reflect utility-specific meter requirements or the performance/warranty claims for a specific make, model, and measurement type.

Many state regulatory bodies have adopted regulations concerning frequency of meter tests, with Table 5-2 showing data from 2015 from various state regulations. Some regulators allow for adjustments in the time intervals between tests when the utility can provide data justifying such a change. While such regulations may not necessarily apply to municipally owned utilities, they often result in better meter maintenance outcomes and are good sources for the initial maintenance program planning. Whether required by a regulatory agency or voluntarily performed by the utility, it must be stressed that meter maintenance is the only means by which revenue is equitably obtained to operate the water system. It requires the necessary time and study to determine the optimum frequency and best practices for testing meters for the most efficient and economic results.

Since meter accuracy degradation can be very gradual, prudent testing intervals, perhaps supplemented by data analytics, can detect significant meter accuracy problems long before large scale meter failure or significant revenue loss occurs. Prudent testing intervals may vary greatly, depending on water quality, usage patterns, and the specific meter types and designs in use in any given utility system. Test interval decisions can also be driven by the rates charged for water (or water and wastewater) service and by water accountability requirements.

Table 5-3 includes recommended data for testing cold-water meters by the use of the volumetric method, using volumetric tanks, gravimetric method, using weight scales, or other calibrated and reliable references. Accuracy standards for new meters are contained in the latest editions of the following AWWA standards: C700, C701, C702, C703, C704, C708, C710, C712, C713, C714, and C715.

The following list provides additional guidance on meter testing beyond the information provided in Table 5-3.

- In testing meters to validate their accuracy either before deployment or after some time in service, note that the testing procedures for electronic meters—which include fluidic-oscillator meters, electromagnetic meters (often called magmeters), and ultrasonic meters—may be different from those meters previously placed in service. The manufacturer's recommended testing procedures should be followed. Ignoring these recommendations can invalidate the accuracy results.
- Electronic meters do not have moving parts and therefore are not subject to the same wear and tear as mechanical meters. Although AWWA standards for mechanical water meters provide recommended testing frequencies in the appendix of each standard, recommended testing frequencies are not provided in the appendix of ANSI/AWWA C715. Water quality and other local conditions may affect the performance of electronic meters over time. To assess these site-specific effects, the utility may choose to verify the accuracy of electronic meters at the recommended test frequencies provided in AWWA standards for mechanical meters until the utility establishes its own criteria based on data collected regarding its operational experience and local conditions.
- Some of the smaller electronic meters have minimum rate accuracy requirements at lower flow rates than any of the rates given in other AWWA meter standards (i.e., flow rates lower than 0.25 gpm). In other instances, the minimum rate flow rates of electronic meters may be lower than those of the mechanical meters previously deployed by the utility for any line size. Accuracy at these lower flow rates, therefore, might be more difficult to test and verify on the existing utility's meter test bench equipment. Gravimetric systems might provide more sensitivity

than volumetric systems. In all cases, the testing equipment must be maintained properly, calibrated periodically, and certified to ensure it is working properly.

- As discussed in the previous section, a utility's testing equipment may not be capable of testing electronic meters reliably at the lower flow rates, particularly for flow rates even lower than the minimums given in ANSI/AWWA C715 if part of the manufacturer's warrantied accuracy claims. If this is the case, an independent third-party service company with certified and calibrated equipment along with sound testing protocols may need to be engaged to verify accuracy. As provided in one of the footnotes for the modified Table 5-3, a utility may alternatively increase the minimum test rate flow to stay within the capability of their equipment while holding the electronic meters to their tighter accuracy requirement at these somewhat higher flow rates. Utilities may also wish to use these somewhat higher flow rates when there are test time constraints.
- Historically, Table 5-3 has included some caveats in test quantity recommendations relative to the footnote that has called for "not less than 3 min running." This has been most notable for some of the maximum rate test quantities that would result in test run times well below 3 min in length. This ambiguity recognizes limitations in the number and size of test tanks with which the utility may be equipped. The footnote further provides that, for reduced test quantities, "the resultant increase in total test uncertainties and errors needs to be recognized." While mechanical water meters may have been able to accommodate those test quantities in Table 5-3 that result in test times of less than 3 min, this may not be the case for some electronic meters. A special test mode—temporarily increasing the sampling rate for the meter—may be available for some electronic meter designs to mitigate the test uncertainties associated with test times shorter than 3 min.
- Table 5-3 provides recommended test flow rates. These recommendations are based in part on the observation that all meters of the same type (all positive displacement meters, all turbine meters, and so on) will exhibit the same or similar characteristic accuracy curves. This may not be a valid observation for electronic meters, for which distinct calibration algorithms may be in use. As a result, alternate or additional flow test rates may be considered. These would be especially relevant for flow rates given in the manufacturer's performance claims that significantly exceed the ranges established in AWWA water meter standards.
- Electromagnetic and ultrasonic electronic meters are inferential in nature, measuring water velocities and inferring volumetric flows based on these velocities. (Many mechanical flow meters are also inferential in nature.) Inferential meters can be susceptible to flow profile disturbances and to "noisy" flow delivery systems (e.g., low frequency pulsations from diaphragm or piston type system pumps or the very high frequency pulsations due to the use of unmitigated variable frequency drives [VFDs] test stand pump drives). Care should be taken to ensure that utility test stands do not introduce disturbances or noise at levels that are not representative of those that occur in actual field installations.

Elements of a Meter Test

Accuracy denotes the comparison between a meter's indicated quantity of water passing through it as compared to the actual quantity of water passing through it as determined by a calibrated (and preferably certified) reference. In the water industry, the meter's accuracy is most commonly called the *percent registration* (% registration). A water meter that has 101-percent accuracy, a percent registration of 101 percent, at a given flow rate indicates that 1 percent more water is registered by the meter than is actually delivered. This error is called 1 percent fast in that the meter over-registers by 1 percent. Conversely, a meter that has 99-percent accuracy indicates that 1 percent less water is registered by the meter than is actually delivered—i.e., 1 percent slow in that the meter under-registers by 1 percent.

Flow range spans the maximum to minimum flow rates established by the manufacturer of the meter being tested. This performance range is sometimes quantified through the ratio of the maximum flow to the minimum flow rates (a ratio also referred to as meter turndown). A meter capable of operating over a range from 20 gpm (4.54 m^3/h) to 0.25 gpm (0.57 m^3/h) has an 80:1 flow range. The flow rates listed in Table 5-3 attempt to standardize testing to best represent meter performance across the entire flow range.

A meter's accuracy curve (also called a performance curve) is a continuous curve drawn between all accuracy points over the meter's flow range. Most water meter manufacturers' literature contains a performance curve for their particular meter model. Generally speaking, all meters of the same general type of mechanical metering technology (disc, piston, multijet, singlejet, compound, or turbine) will exhibit the same or similar characteristic accuracy curve. Electronic meter accuracy curves, in contrast, may be dissimilar, as specific calibration algorithms can lead to distinct curves.

The following are the basic elements to be determined for any meter test.

- The specific flow rates at which to test, within the operating range of the meter
- The quantity of water (i.e., the test draft) to use at each of the test flow rates. (Smaller draft sizes provide poor reading resolution and can increase other test uncertainties, while very large draft sizes may not be practical.)
- Acceptance limits at each of the test flow rates
- The order in which the flow rates are tested

With regard to the last item, acceptance testing should be the highest to the lowest flow rates, while in-service and out-test are lowest flow rates to the highest. Electronic meters might have varying requirements, and utilities should review the manufacturer's recommended procedures.

In testing meters to validate their accuracy (either before deployment or after some time in service), note that the testing procedures of electronic meters may be different from those meters previously placed in service. The manufacturer's recommended testing procedures should be followed. Ignoring these recommendations can invalidate accuracy results.

Accuracy Issues

Individual accuracy test results need to be assessed within context of possible test errors.

Repeatability is the measure of deviation of a series of accuracy tests from the test's mean value where all of these accuracy tests were conducted under exactly identical conditions. Deviations in test results are driven by individual components within the test system—including the operator—as well as the meter under test. With poor repeatability, the average of multiple tests could indicate good meter accuracy, while any individual test might indicate poor accuracy.

Precision usually relates to the resolution to which individual measurements can be read. Poor precision will inherently produce poor repeatability. If a gravimetric reference for 100 gal (834 lb) can be read to within 0.1 lb, the precision of this reading is quite high (to within 0.012 percent of the 834-lb reading). But the repeatability of the scale reading, if one imposes the same set of National Institute of Standards and Technology (NIST)–traceable weights on the scale multiple times, might be found to be on the order of a 0.3 lb (for example). In addition, the scale may be found to have a systematic error, relative to these traceable weights, of 0.5 lb (for example), based upon the average of multiple weighings. So in this gravimetric example, the scale reading is much more precise than it is accurate.

Uncertainty usually refers to the combined sources of all errors (commonly termed systematic and random) in a test system which may include data presentation (such as precision limitations) and operator interpretation errors as well as the combined proving system repeatability errors of each measuring component and measurement. In addition to these sources of error, a test system might have a bias (i.e., calibration error or systematic error) where the test system-indicated proof quantity is displaced from the true value at one or all proof quantities.

Test Rates

Per the recommendations in Table 5-3, the three test flow rates used to properly assess the performance of displacement, compound, propeller, and other types of water meters are maximum, intermediate, and minimum. All three rates are necessary to evaluate the overall meter performance. At least one additional test, preferably more, is necessary within the changeover range of flows of compound and compound-type fire-service meters to determine overall operational efficiency and accuracy of registration. These recommendations are based upon general trends in meter accuracy curves. Other test flow rates may be added, as driven by manufacturers' claims, individual utility specifications, application-specific concerns, and other factors. Actual rate-of-flow at each test point needs to be measured. Rates that are instead based solely on fixed orifices are problematic since uncertainties on orifice size and positioning, as well as variations in delivery pressure, can greatly affect flow rate.

Tests for full-flow (maximum test flow) accuracy do not need to be made at the "safe maximum capacity" rate shown in the applicable AWWA standard. Registration curves of water meters show that meters in good operating condition follow a general pattern of registration. Typically, accuracy is fairly constant at rates well above the intermediate rate of flow, as noted in the next paragraph. Selection of the maximum rate of flow at which meters are tested is therefore not of major importance. Maximum-rate test flows of 25 percent or more, if desired by the owner, of rated capacity are practical because meters are seldom operated at rated capacity. This lower test rate is advantageous with multiple testing of small meters and is possible with pressures and testing equipment usually available. There might be piping restrictions for field testing or tank size limitations affecting the maximum rate that can be tested.

The intermediate rate of flow is normally intended to be at or near the high point of registration to ensure against over-registration of the meter on any rate of flow. It is often found to be at approximately 10 percent of rated maximum flow rate. This may not apply for some meter types. An examination of the manufacturer's accuracy curve technical data might identify flow rate regions where accuracy approaches the acceptance limits (high or low) that may be worth some level of assessment.

The minimum-flow test discloses operational ability and proficiency of meter repair more than either the maximum- or intermediate-flow tests. In many applications, changing consumer usage patterns make accuracy at lower flow rates increasingly important in reducing nonrevenue water use. The minimum test rate is normally the "minimum test flow" value from the relevant AWWA meter standard. With the introduction of AWWA C715 meters, however, some of the "minimum test flows" are at flow rates below the established capabilities of some test facilities. Gravimetric systems might provide more sensitivity than volumetric systems. In all cases, test equipment must be maintained properly, calibrated periodically, and certified to ensure it is working properly. These flow rates may also result in extremely long test times if test uncertainties are to be kept at acceptable levels. As a result, a footnote in Table 5-3 notes that an alternative test at the lowest flow within the "normal test flow limits" with attendant tighter accuracy limits may be a viable substitute to reduce test times or to stay within the capabilities of the test equipment.

If accuracy tests are being conducted for warranty purposes, the utility might consider testing only at the low flow rate (and perhaps the intermediate test rate) if failures are seen at these rates. Refraining from testing at the high flow rate may preserve measuring element conditions that have caused these test failures. The utility may also consider requesting recommended test procedures from the meter manufacturer to avoid conflicting test results.

Test Volume

The quantity of water required to provide acceptable accuracy depends on the accuracy and resolution of the test equipment, the reading resolution and other metrological characteristics of the meter under test, and the desired accuracy and repeatability of the test results. Calibrated volumetric tanks with adjustable read height scales can be read with an uncertainty of less than ±0.25 percent if the scale reading error is small compared to the final height of the collected water. Measuring tanks based on weight, referred to as the gravimetric method, can be read within even tighter uncertainties (perhaps as low as ±0.10 percent) when using Class II or Class III precision scales, load cells, and sufficient test drafts/volumes. While reference meters may result in larger uncertainties than volumetric or gravimetric references for the same test quantity, such meters provide greater flexibility in allowing the use of increased test quantities. Test quantities may need to be increased in order to provide suitable minimum test times. Refer to both Table 5-3 and manufacturers' test guidance for recommended test quantities/times.

Since repeatability will vary depending on the test equipment, the operator, and the type of meter, it is good practice to run three tests with the same draft size at a given flow rate to assess repeatability. Optimally, the range in the results of these three tests should be within one-fifth of the range in allowable accuracy. For example, three tests at the maximum flow rate, or Qmax, of 15 gpm for a C700 ⅝ in. should have a maximum range of 0.6 percent, while a ¼ in. gpm should have a maximum range of 1.3 percent. This is a rough measure of the random uncertainty component within the total test uncertainty. If repeatability is found to exceed this desired level, one option is to increase the test draft or test time (perhaps doubling, as a first attempt). If longer test times are not possible or practical, the identified repeatability should be considered when assessing accuracy test results.

Test Automation, Test Compression

Where large quantities of water are used for testing large meters, it is more expedient to employ gravimetric means involving the use of computers and electronics with load cells. These procedures make possible testing "on the fly," thereby eliminating the effects of valve opening and closing in supplying the test flow rate, particularly at high flows, and allowing adequate test accuracy even with compressed test quantities, as long as adequate reading resolution is possible for the meter under test. Such on-the-fly testing at higher flow rates typically requires the automated operation of calibrated diversion valves.

Table 5-4 is intended for ascertaining the percentage of true quantity indicated by the meter when the meter is run for a measured tank volume of 20, 30, 40, 50, or 60 gal in examples of such compressed test quantities. Limits in reading resolution will result in larger test uncertainties for compressed test quantities. Compressed quantities when using volumetric references can result in significant errors and are not recommended.

Table 5-4 Percentage registration tables for test quantities other than 10,100 or 1,000 gal or ft³

Meter Reading, *ft³ or gal*	Percentage of Actual Volume	Meter Reading, *ft³ or gal*	Percentage of Actual Volume	Meter Reading, *ft³ or gal*	Percentage of Actual Volume	Meter Reading, *ft³ or gal*	Percentage of Actual Volume	Meter Reading, *ft³ or gal*	Percentage of Actual Volume
19.0	95.0	29.0	96.6	39.0	97.5	49.0	98.0	59.0	98.3
1	95.5	1	97.0	1	97.7	1	98.2	1	98.5
2	96.0	2	97.3	2	98.0	2	98.4	2	98.6
3	96.5	3	97.6	3	98.2	3	98.6	3	98.8
4	97.0	4	98.0	4	98.5	4	98.8	4	99.0
5	97.5	5	98.3	5	98.7	5	99.0	5	99.1
6	98.0	6	98.6	6	99.0	6	99.2	6	99.3
7	98.5	7	99.0	7	99.2	7	99.4	7	99.5
8	99.0	8	99.3	8	99.5	8	99.6	8	99.6
19.9	99.5	9	99.6	9	99.7	9	99.8	9	99.8
				Actual Volume*					
20.0	100.0	30.0	100.0	40.0	100.0	50.0	100.0	60.0	100.0
1	100.5	1	100.3	1	100.2	1	100.2	1	100.1
2	101.0	2	100.6	2	100.5	2	100.4	2	100.3
3	101.5	3	101.0	3	100.7	3	100.6	3	100.5
4	102.0	4	101.3	4	101.0	4	100.8	4	100.6
5	102.5	5	101.6	5	101.2	5	101.0	5	100.8
6	103.0	6	102.0	6	101.5	6	101.2	6	101.0
7	103.5	7	102.3	7	101.7	7	101.4	7	101.1
8	104.0	8	102.6	8	102.0	8	101.6	8	101.3
20.9	104.5	9	103.0	9	102.2	9	101.8	9	101.5
21.0	105.0	31.0	103.3	41.0	102.5	51.0	102.0	61.0	101.6

*For 100 division dial, move decimal one place to right.
NOTE: Metric conversion: gal × 0.003785 = m³.

Using automated testing equipment can also free operators to perform other tasks while tests are being run. For example, simply automating a shut-off at test's end can prevent tank overflows.

Testing Interval

Ongoing meter-testing programs have long been advocated by AWWA and are required by most public utility commissions and other regulatory agencies (see Table 5-2). It is in the best interest of both the utility and the customer that testing of meters be part of an ongoing maintenance program.

The characteristics of the water entering a meter can greatly affect meter performance. This is one reason that an arbitrary number of years is not the criterion to use for determining the length of time between tests. From an economic standpoint, a meter should remain in service until its level of under-registration justifies its replacement. Because meter testing programs are costly, prudent management for revenue protection dictates that meters should be left in service as long as practical. Because of these variable factors, it is recommended that a utility's own test results be used to determine the length of time its meters should remain in service between tests. The utility may also consider using test intervals that are recommended for use in their area. Analytics of specifically designed data—such as hourly meter readings or accuracy testing data constructed by meter brand, size, and type—could also be used to provide economic justification for test interval decisions.

Statistical sampling can allow a utility to establish its own periodic meter-test intervals. These should be flexible so that different test intervals may be established for different types of meters or for different manufacturers' meters. Different time intervals may also be established for meters in areas supplied with water from different sources if those sources result in water quality variances. For water utilities where meter readings are used for wastewater service billing, the impact of meter accuracy on that revenue should also be considered when establishing testing intervals.

Electronic meters do not have moving parts and are therefore not subject to the same wear and tear as mechanical meters. So while AWWA standards for mechanical water meters provide recommended testing frequencies (in the appendix of each standard), no testing frequencies are provided in the appendix of C715. Water quality and other local conditions, however, may affect the performance of electronic meters over time. To assess these site-specific effects, utilities may wish to verify the accuracy of electronic meters at the test frequencies provided in AWWA standards for mechanical meters until the utility establishes its own criteria based on data collected on its operational experience and local conditions.

Statistical Sample Testing

Sample testing is an alternative method to evaluate the performance and service life of a cross section of meters and for the auditing of new meter shipments. In addition to providing insight into variables affecting meter performance, this method also allows the utility to monitor the overall accuracy of its meters. Sampling is a useful management tool in addressing metering activities, as well as identifying problems with specific meters.

Properly designed statistical sample testing identifies the poorest performing meters and allows the utility to exchange meters based on performance rather than an on arbitrary time-in-service threshold or one based on cumulative throughput.

The first steps to implement a sample testing program are to understand and identify the established statistical methods. This may be accomplished through educational training or by referencing basic statistical methods in textbooks.

After developing a basic understanding of statistical sampling, the following steps should be implemented.

1. Determine desired confidence level.
2. Determine appropriate sample lot size for the confidence level.
3. Determine criteria for testing (size, age, volume, type, and so on).
4. Randomly select and retrieve.
5. Test and document.
6. Analyze test data.
7. Report with recommendations.

The following is an example that identifies the benefits of sample testing and performance-based meter changeouts versus annual changeout based on age. In this example, a 95-percent confidence level is targeted. The appropriate lot size to achieve a 95-percent confidence level was determined to be a minimum of 5 percent of the total meters for a particular set year, regardless of total population size. The criteria used in this example to sample test meters was the year in which the meter was put in service or the "set year." A minimum of 5 percent of the meters from each sample lot (set year) is randomly pulled from field service and is returned to the meter test facility for shop testing.

Testing standards identified in this manual should be followed and adhered to. Extreme care should be used when retrieving and testing sample metering to best duplicate field accuracy/conditions and to prevent damage to the meters. The data in the example include an overall weighted average determined by assigning weights of 15, 70, and 15 percent to the low, intermediate, and high test results, respectively (Tao 1982). Based upon recent residential end-use studies, other weighing functions that place more importance on low flows than this 15–70–15 model are also now in use.

Once the average accuracy of the sample lot is determined, the utility has several options to consider. If the sample lot meets or exceeds the utility's accuracy requirements and therefore would not be economic for replacement, this result indicates that the meters represented by the sample lot can remain in service until the next time this lot is selected for performance testing. The meters that are tested could also be redeployed. If the accuracy test results are borderline for replacement, the utility could conduct another sample size test to confirm results or perhaps conduct another weighted average calculation using different percentages of low, intermediate, and high flow rates to determine if that helps inform a replacement decision. Any such changes in weighting should be based on usage profiling from datalogging or other credible and documented sources to support the result. Since conducting another sample lot test is expensive, a desktop analysis might be preferred to obtaining meters from the field. Lastly, if the average accuracy of the sample lot indicates that it would be economic to replace the entire meter population represented by the sample lot, the utility might wish to consider conducting another sample to confirm results before justifying and scheduling a replacement. Organizing and retaining these sample test results over time are valuable tools for making meter procurement and application decisions.

Test Equipment

For both dedicated test facilities and for portable test equipment used for in-situ testing, test equipment must be properly constructed, maintained, and calibrated. Calibration should be done on an annual basis at a minimum and as required following repairs or modifications. Calibration should be done against traceable standards and documented.

Equipment required to test a water meter may be very simple (Figure 5-2). A rotameter is a useful flow-measuring device for precise control in meter testing. It consists of a tapered, calibrated glass tube in which a stainless steel rotor is free to move up and down in the center of the tube guided by a stainless steel guide. The instrument is positioned above a flow-regulating valve. Water passing up the tapered tube raises the rotor to a stable position. The rate of flow is read across the top of the rotor on a scale of figures etched in the body of the glass tube. Rotameters are commercially available for meter repair–shop testing apparatus. Figure 5-3 illustrates basic features of a rotameter. While Figure 5-2 shows a pressure gauge upstream of the meter(s) under test, a relocated or additional gauge downstream of the meters may better monitor the risk of cavitation created by high flow rates coupled with low pressures. (Cavitation can be further exacerbated by elevated water temperatures and/or high dissolved gas content.)

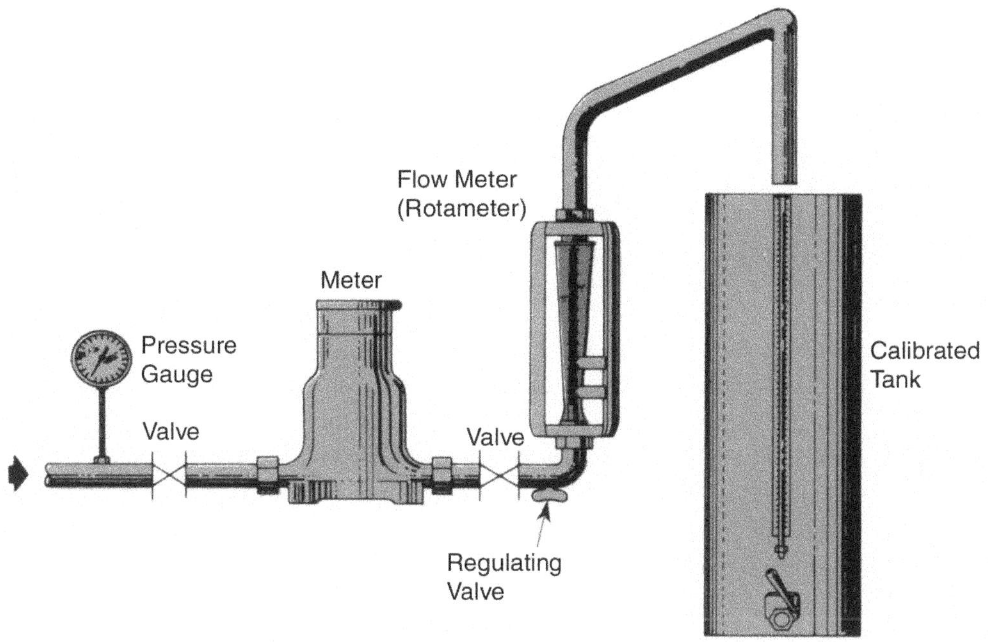

Figure 5-2 Basic requirements for a volumetric meter-testing assembly

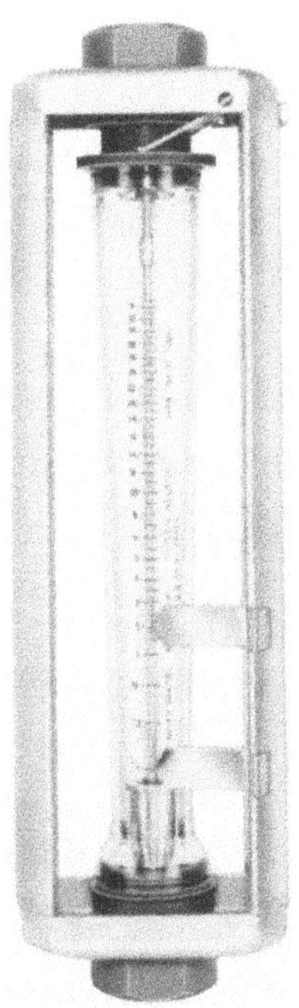

Figure 5-3 Rotameter rate-of-flow indicator

At high flow rates, modern scale systems have built-in buffers that are programmable to eliminate the effect of impact and vibrations and produce accurate flow data. While scale systems can also deliver average flow data for the entire duration of the test, good operating practice suggests comparing those values with the actual timed flow rates for verification of the built-in buffers. Under "quiet" discharge conditions into a gravimetric tank, the rate of change in indicated tank weight can be used instead to measure the flow rate being delivered during the test.

Fully assembled, "turn-key" test equipment is commercially available. Also, the various meter-testing components may be purchased individually from commercial manufacturers, and an organization can fabricate its own test equipment for any particular application. Meters may be tested singly or in groups, and the equipment selected should be based on the workload. When reasonable judgment is exercised in the selection of test equipment, the cost is quickly repaid and more accurate results are usually obtained. Requisites of a good meter test bench and facility include the following.

- A stable test bench on which meters may be quickly and securely held, in proper alignment
- Inlet and outlet piping at the bench should provide undisturbed flow profiles when testing meters that may be susceptible to flow profile disturbance.
- An inlet valve
- A quick-closing valve on bench discharge and a flow-regulating valve
- Intermediate coupling pieces (spools, idlers, adapters) to be placed between meters for those designs that may be susceptible to upstream or downstream flow profile disturbances
- A rate-of-flow indicator
- One or more tanks, preferably calibrated for volumetric testing, or a tank set on platform scales for testing by weight
- Calibrated tanks installed in a plumb position to avoid inaccuracies of calibration
- Adequate lighting of test area
- Cabinets for orderly storage of repair parts, hand tools, and records
- A recirculating system or an ample supply of water so that pressure and flow rate will fluctuate as little as possible
- Constant, nonpulsating flow conditions may be especially important for some meter types like velocity meters.
- Convenient and accessible location and carefully designed shop layout to increase efficiency
- Floor drains to handle tank and bench overflows
- Lifting equipment to move larger meters and associated spool pieces
- Racks for storing spool pieces used in testing larger meters
- Provisions for measuring water temperature

Procedures for consistent test practices and leak testing as part of in-process checks are important to maintain a good test bench facility. Running water into the tank before accuracy testing is also a good test practice for consistent results.

Safety is an important consideration because water in motion can be very dangerous. Certified equipment should be used for accuracy and technician safety.

Basic equipment for testing ⅝-in. (15-mm), ¾-in. (20-mm), and 1-in. (25-mm) mechanical meters consists of the following items.

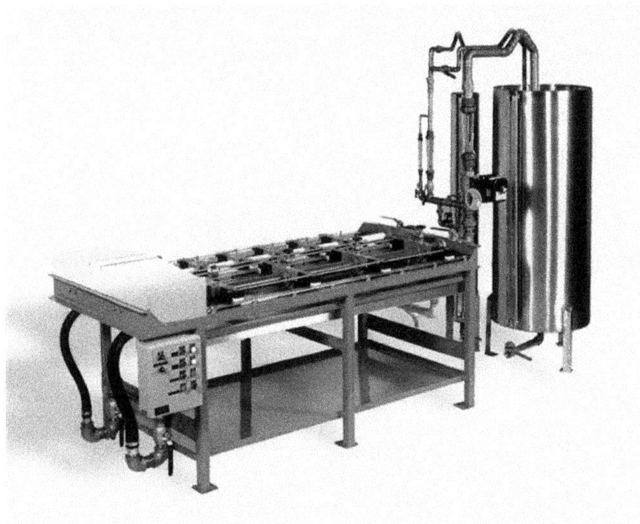

This test bench can test ⅝-in. (15-mm), ⅝-in. x ¾-in. (15-mm x 20-mm), or 1-in. (25-mm) meters. The unit features a rate-of-flow indicator and an electronically actuated shutoff valve, with two calibrated tanks. This style bench features mechanically operated clamping mechanisms.

Figure 5-4 Test bench for small meters

- A 10-ft^3 or 100-gal (0.5-m^3) capacity tank, calibrated for volumetric testing, or fitted with gravimetric scales
- A 1-ft^3 or 10-gal (0.05-m^3) capacity tank, similarly equipped

  (As an alternative to volumetric or gravimetric references, reference meters are sometimes used. These provide greater flexibility in the test draft size selection but can introduce additional test uncertainties: The reference meter repeatability, calibration/accuracy, and resolution must all be assessed.)

- A test bench furnished with necessary fittings, including a control inlet valve, a quick-acting rate control valve on the discharge, and a rate-of-flow indicator

The bench may be of single-meter capacity or designed for simultaneous testing of a number of meters (Figure 5-4). The test bench in Figure 5-4 can test ⅝-in. (15-mm), ⅝-in. × ¾-in. (15-mm × 20-mm), or in 1-in. (25-mm) meters. This unit features a rate-of-flow indicator and an electronically actuated shutoff valve with two calibrated tanks. This style of bench features mechanically operated clamping mechanisms.

Similar equipment should be provided for testing meters larger than 1 in. An additional tank of 100 ft^3 or 1,000 gal (5 m^3) or larger should be provided for testing 3-in. (80-mm) and larger meters. Commercial testing is available for all meter sizes used in modern utilities, including fire-line meters (see Figures 5-5 and 5-6). A rate-of-flow indicator of larger capacity is also essential. The test bench in Figure 5-5 can test 3-in. (80-mm) through 12-in. (300-mm) water meters of different types. Multiple meters can be tested depending on their size and length. The test facility in Figure 5-6 includes two 50-hp pumps rated at 1,250 gpm (284 m^3/h) each. It includes two 5,000-gal or 500-ft^3 (19-m^3) tanks in addition to several smaller tanks, all based on gravimetric systems. A hydraulic system is used to clamp the meters in the place, and the bench is operated using a programmable gravimetric system.

When testing meters pulled from service in the field, there may be risk of debris breaking loose and affecting items downstream of the meter under test. It may be useful to install a strainer or other protective measure to protect the test equipment from such debris.

This test bench can test 3-in. (80-mm) through 12-in. (300-mm) water meters of different types. Multiple meters can be tested depending on their size and length. A hydraulic system is used to clamp the meters in place, and the bench is operated using a programmable gravimetric system. All water is recirculated.

Figure 5-5 Testing of large meters

This test facility includes two 50-hp pumps rated at 1,250 gpm (284 m^3/h) each. It includes two 5,000-gal or 500-ft^3 (19-m^3) tanks in addition to several smaller tanks, all based on gravimetric systems.

Figure 5-6 Large meter testing facility

Procedure

The 13-step description in this section covers the actual steps for acceptance and postrepair testing of a single ⅝-in. (15-mm) mechanical meter in a high-to-low flow sequence (see step 11 for the testing of in-service meters); the only differences for larger meters are the rates of flow and test quantities used. Test equipment is the same as that previously described. Test practices may be slightly different for electronic meters; the user should refer to the manufacturer's recommendations for testing, based upon considerations such as totalizer update rates, reverse-flow registry and so forth. For meters pulled from service in the field, care should be taken to ensure that

- no foreign debris or loosened scales were introduced as the meter was removed from the line and transported to the test facility;
- internal surfaces were kept hydrated with fresh water or surfaces were rehydrated before test (if it can be determined that the drying-and-rehydration sequence did not loosen any scales or damaged any seals); and
- no mechanical damage occurred during removal or subsequent handling of the meter.

1. Secure the meter in the test bench. (Do not tighten more than necessary to make a watertight connection, as there is a possibility of distorting the meter housing or extruding the washer into the waterway.)
2. Open the register cover.
3. Open the discharge valve first, then open the inlet valve gradually and run the water to waste until the entrapped air is cleared. This process also ensures a full discharge line to the tank. In the case of acceptance testing of new meters, the highest flow rate achieved during this air purge should be as great as that intended for the maximum-flow test (see point 7). This air purge may take as little as a few seconds or may require half a minute or more (as this can depend on the type of stand, the type of meter, and the number of units being tested at one time). A clear sight tube downstream of the meters under test (or the clear float tube of a rotameter) can provide visual indication of continued air passage, prompting additional air purge. Manufacturers may be able to provide specific purging requirements for their meters.

   Note step 11 for determining the condition of the meter as removed from service in the field. In this instance, purging at flows that ramp up to maximum-flow test rate may not be correct if such flows were rarely, if ever, experienced in the field. For such considerations, the maximum purging flow rate should be comparable to the maximum rate anticipated at the field site, with purging times increased accordingly. Roughly speaking, this flow might range between one-third and two-thirds of the maximum-flow test rate. A utility may wish to consider the reason for removal from field service to determine the purge rate as to not impact the performance as removed. Two risks or sources of possible test error need to be balanced: If air has not been fully purged from the meter, results will not be representative of actual field performance, while if the purge practices use excessive flow rates, foreign material on bearing or metering surfaces—that may have affected meter performance while in the field—might be scoured away before the accuracy tests are performed.

   Note that some test benches (and even some meters or meter assemblies) include small bleed ports that can be used to remove some of this entrapped air. If present, these features must be monitored to prevent test errors due to leaks.

4. Shut off the discharge valve.
5. Check 100-gal or 10-ft^3 (0.5-m^3) tank discharge to ensure that the tank is empty and then close the tank drain valve. (Tank discharge should be to an open drain so that any possible leakage of the tank drain valve can also be observed.) Tare/zero out the weight scale if running a gravimetric test.
6. Record the correct meter reading carefully. The purge cycle procedure takes up any possible backlash in the gearing of mechanical meters. This is often simplified by using a bleed port downstream of the meter as part of the purge process in step 3 to "zero-out" the meter under test.
7. Open the test bench discharge valve as rapidly as prudently possible to the desired rate for the maximum-flow test (15 gpm [3.4 m^3/h]). Continue the flow at this rate until the meter reaches the desired quantity, with a smooth valve closure, avoiding a too-rapid closure that may cause water hammer and/or a "jump" in mechanical registration gear backlash. Read the meter accuracy from the scale of the volumetrically calibrated tank (Figure 5-7) or by weighing the water in the tank. The calibration scale in Figure 5-7 shows the percentage by which the tested meter is over- or under-registering. The reading illustrated is 0.5 percent faster because the meter registers 100.5 percent of true quantity passing through it. Trickling or bumping water through the meter to accomplish an exact volume of measurement is not recommended. If the indicated test volume shown by the meter under test is not exactly equal to the nominal test quantity that had been intended, the meter accuracy calculation must include consideration of the actual volume indicated by the test meter (see similar considerations under the section on "Multiple-Meter Testing" that follows).

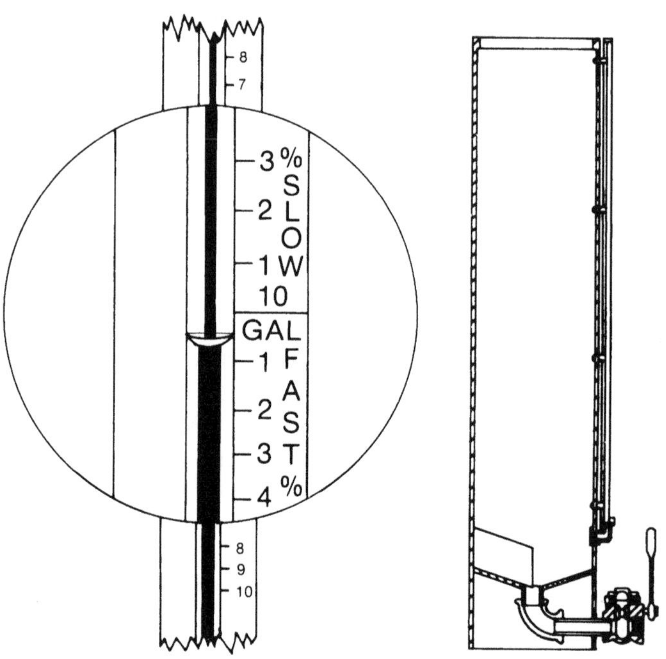

The calibration scale shows the amount by which the tested meter is over- or under-registering, in percent. The reading illustrated is 0.5 percent fast (meter registers 100.5 percent of true quantity passed through it).

Figure 5-7 Detail of calibration scale of testing tank and sectional view of tank with outlet valve

8. Record the results of this test on the record form or log data into the computers used for meter management.
9. If the meter being tested is a repaired meter and its accuracy does not fall within prescribed limits, use calibration adjustments or new parts to bring the unit within limits. Repeat the test as outlined. Be sure that the meter tests correctly by rerunning the test after any changes are made.
10. If a repaired meter is being tested and is found to register within the limits on the maximum-flow test rate, repeat steps 3 through 8 for the intermediate (2 gpm [0.45 m^3/h]) and minimum (¼ gpm [0.057 m^3/h]) tests, discharging the test water into the 10-gal or 1-ft^3 [0.05-m^3] tank. The test quantity for these rates is only 10 gal or 1 ft^3 (0.05 m^3), and the meter register test hand completes only one full revolution.
11. If, however, the test is to determine the condition of a meter removed from service, it is recommended to rearrange the order of test flows. The preferable order in this event is the minimum-, intermediate-, and, finally, the maximum-flow test. If a meter that has been in service for some length of time is tested first on the high rate of flow, there is a possibility of freeing the disc and thereby obtaining a false impression of the meter's condition on lower rates. Conversely, a meter should be tested as soon as possible after removal from service to prevent the drying of deposits in the measuring chamber, as this condition tends to give an adverse impression of the meter's condition while it was in service. Note that there is a risk that the actions taken in removing the meter from the field setting have introduced debris into the measuring chamber, so that the in-service condition of the meter no longer exists. In addition, for multi-meter testing (see below), there is a concern that debris originating in one meter may move during testing and affect downstream meters.

    (If a meter exhibits poor accuracies at the minimum or intermediate rates under this rearranged order of test flows, there may be no way to determine if this represents actual performance in the field. Continuing the test sequence by progressing to the maximum rate test might then be problematic. These higher velocities might exceed those previously experienced in the field, and so change conditions within the meter through scouring or some type of "break-in" behavior. On the other hand, the higher velocities might purge debris that was introduced into the meters during removal or transport, as well as any remaining entrapped air. If the maximum recurring flow rate actually experienced in the field is known, this rate may be a suitable substitute for the maximum test rate provided in Table 5-3.)
12. After the three separate tests have been run, the meter has been fully tested and may be removed from the test bench after the inlet valve of the test bench has been closed and pressure has been released by a partial opening of the discharge valve.
13. Once the test is concluded, the meter should be drained, the register sealed, and the dust caps placed on the meter spuds before the meters are put in storage.

The calibration of a tank is based on its rated capacity, and if lesser quantities are used in testing, as happens sometimes, the calibrated percentage errors are not valid.

It should be noted that a rate-of-flow indicator may be affected by water temperature at lower flow rates. Most rate-of-flow indicators are calibrated at a specific temperature, and this calibration is affected by temperature-driven changes in density or viscosity. For rotameters, the indicated rate-of-flow typically increases for low-temperature water and decreases for high-temperature water. Eq 5-1 may be used to verify the actual rate of flow during a standard accuracy test:

$$RT = V \qquad \text{(Eq 5-1)}$$

Where:

R = rate of flow, gpm
T = time of flow, min
V = volume of test tank, gal

Example: It takes 4 min to fill a 1-gal calibrated cylinder at ¼ gpm:

$$R = \frac{V}{T} = \frac{1}{4} = \frac{1}{4} \text{ gpm}$$

This check on the accuracy of low-flow calibration should be considered when a marked change in water temperature of the test water occurs, or the entire test volume should be timed for the actual flow rate.

Any leaks (other than ambient leaks immediately upstream of the first—or only—meter under test) will affect test results. Similarly, test errors will be introduced by improper draining of test tanks or by interrupted draining of the lines that discharge into the test tanks.

Multiple-Meter Testing

Multiple testing of meters is identical to the testing of one meter, except for one important factor. In multiple-meter testing, it is not possible to stop the test when the meters under test reach the exact test quantity that is targeted, since each meter is likely to indicate a slightly different volumetric throughput over the course of the test. So instead of using a reference tank reading of "X" percent fast or slow versus the meter under test indicating the intended test quantity, the test is stopped when the reference tank or calibrated meter reaches the intended test quantity. The individual registration reading of each meter under test must then be assessed in turn.

In mentally calculating the accuracy of each meter in turn, note that the significance of each graduation of the test index (with respect to percentage accuracy) depends upon the test quantity. For example, a ⅝-in. meter may have a 10-gal test index divided into 100 graduations. For a 10-gal test volume, each graduation represents a 1-percent change in registration. For a 5-gal test volume, each graduation instead represents a 2-percent change in registration.

If the delivered test volume/weight in the reference tank—or calibrated meter—is not exactly equal to the targeted quantity, the calculated accuracy of each meter must be adjusted accordingly. Again, trickling or "bumping" water through the test line until the targeted test quantity is reached on the reference is not recommended.

Although concern is often expressed over the time involved in testing meters on low rates of flow and large test quantities, multiple testing of meters reduces this factor on a unit basis. Furthermore, a relatively low-cost, automatic, electric cutoff valve can be installed in the test bench unit, which eliminates the need for someone to stand by to ensure against overrun of the test. This attachment, shown in Figure 5-8, consists of an electric valve with a strainer, a relay with transformer, one low-voltage electrode for each tank, and the necessary wiring and switches. When the water in the tank rises to the test volume level, this causes the relay contacts to open and the valve to close and stop the flow. Gravimetric systems are usually supplied with automatic shut-off for greater accuracy. (See Figure 5-9 for a large multiple-testing operation that can test ⅝ and ¾ meters in production quantities.)

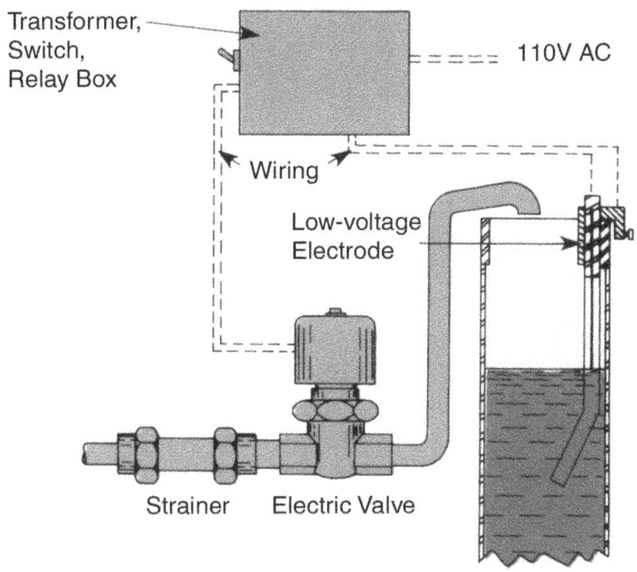

This device automatically stops the flow of water when the testing tank contains the preset quantity of water. This feature eliminates the need for someone to stand watch to ensure against overrun of the text.

Figure 5-8 Automatic cutoff valve

This bench can test ⅝ in. and ¾ in. meters in production quantities. Two rows at a time are tested at high, medium, and low flow rates.

Figure 5-9 Large multiple-meter test bench

Large Meter Considerations

Larger meters are sometimes ignored as long as they continue to record consumption. These meters may be few in number, but they represent a significant amount of revenue for a water system. Often, the largest 10 percent of the meters measure 40–60 percent of a system's consumption. Revenue potential at any specific account should be considered when prioritizing the meters to be tested.

Considering the fact that larger meters are so important to a water system's financial health, they should be maintained to provide peak performance. Maintaining older meters can be challenging since they are harder to work on and repair, spare parts are expensive or not readily available, assemblies are sometimes complex, and it takes more knowledge and better-trained repair personnel to maintain them than for smaller meters. However, current large meter designs—like turbines, compounds, and fire-line meters—are produced in modular designs with consideration for regular maintenance. Electronic meters have no moving parts and might be effective alternatives to address maintenance issues. However, periodic testing of these types of meters may still be prudent.

These larger meters are so important that their operating condition must be monitored on a systematic and timely basis. One alternative approach is for qualified test personnel to test these meters on site. Increased revenue and water accountability gains can offset the initial investment and continuing costs of such a testing program.

Testing Considerations

There are certain advantages to testing the meters at the water system's facilities, including better access to repair equipment and parts, as well as facilities and conditions that may provide better control of test conditions, the range of flow rates tested, and smaller cumulative test errors. Impacts on customer operations and safety are also important considerations in making test location decisions. However, there are situations where onsite testing can be a valuable alternative.

From a technical standpoint, the piping in a meter installation can have a definite influence on a meter's accuracy, and this can be detected by the onsite tests. For example, the diagram in Figure 4-6 details correct meter installations for turbine meters. Properly operating isolation valves and a straight pipe, a minimum of 5 pipe diameters in front of the meter, and a minimum of 3 pipe diameters downstream of the test tee are critical in attaining engineered accuracy and performance. Meter settings that do not comply with installation recommendations may be candidates for onsite testing.

In addition, the act of removing a meter from its field setting and transporting it to a test facility can itself introduce variation in meter behavior. This might be especially true for larger meter sizes that need specialized equipment to remove the meter from service or that take additional staff for safe removal (as in vault applications).

There are often constraints in testing meters on site. If the meter setting does not have a bypass line, the customer will be out of service for the duration of the testing. The location of the meter may provide poor access for personnel, test equipment, and discharge lines. There must be a safe discharge path for the test water. Isolation valves used to divert water during the test may not operate properly due to lack of exercising. Debris or water discoloration generated by the testing may be of concern to the customer. Test port size may limit the maximum test flows that can be attained; this may be important if consumption patterns at the site include significant usage at flows well above those possible in the onsite test.

One acceptable method of maintaining proper performance of certain types of larger meters is to replace the operating components and assemblies while leaving the meter body in place. If the measuring and registration functions are within one integral assembly and shop tested, accuracy tests are not required after installation. If separate assemblies are involved, a final onsite accuracy test is appropriate. Without accuracy testing of separate assemblies, the potential loss of significant revenue increases.

An additional approach that might be useful in challenging situations is the use of a clamp-on ultrasonic meter. Given the possibility of relatively large uncertainties in clamp-on meter accuracies, this approach may have limited, site-specific applications. However, a clamp-on reference meter might be able to indicate that the field meter under test has serious accuracy issues, meriting additional investigation or repair/replace actions. In using clamp-on meters, manufacturer's instructions must be carefully followed with specialized training and routine calibration checks. Understanding the strengths and weaknesses of this approach, a utility may decide to add clamp-on ultrasonic meters to their testing methods.

Field Testing

Setting up and maintaining an onsite meter testing program involves a number of factors. The meter setting (whether indoors or in a vault/pit) must have adequate and safe space in which personnel can maneuver. Occupational Safety and Health Administration requirements must be adhered to, and everyone in the testing program should be clearly instructed on its guidelines. The piping arrangement around the meter must include some method to positively isolate the meter while still maintaining an adequate flow to the end user. Temporary or permanent bypass piping needs to be provided.

Some larger meters have built-in test plugs; others do not. Test ports may not be suitable for some higher flow rates. For installations that require test outlets, the outlets can be fabricated in a number of ways. Service saddles and reducing tees are the most frequently used approaches. These need to be installed according to the recommendations of the meter manufacturer and located so that the connecting hose to the onsite tester is correctly located downstream of the meter and can be properly secured/constrained.

A prerequisite to testing is that the isolating downstream shut-off valve provides positive shut-off. If the valve is not completely operational and there is leakage when the valve is in the off position, the test will not be accurate.

To assist the testing operation, a short length of pipe may be permanently attached to the test outlet, along with a shut-off valve that can be locked into position. This will save considerable time and effort in preparing the meter site for periodic testing.

The personnel assigned to perform the tests must be properly trained and have the appropriate test and safety equipment. There are certain techniques and procedures that must be properly followed when using the test equipment. The consequences of discharging large volumes of water at high flow rates must be understood, appreciated, and accounted for, including concerns over water hammer if the test flows or test valve closures are more extreme than those normally experienced at the site. Some utilities actually capture test water for conservation and use elsewhere. Improper use of the equipment can be harmful to the operators, the meter, and the surrounding area.

A comprehensive site survey can help determine if an onsite test can be done safely and to full effect. This information can help both testing personnel and the staff responsible for meter replacement. This survey should include an assessment of shut-off valve status (perhaps including work with the customer's staff to assess the customer's shut-off valves as well). This can prevent sending crews to sites where the testing cannot be successfully performed.

Finally, the water customer must be prepared for the testing. They may need to provide access to the test crew and may be warned to expect temporary service interruption, reduction in supply pressure, entrainment of debris into the flow stream, etc. Often, if a meter testing program is well established, water customers can work with the meter-testing crews to minimize the disruption caused by the testing.

The techniques for performing the tests, selecting the appropriate test flow rates, determining the accuracies, and reaching conclusions must be known and carefully followed to obtain valid test results. Meter testers should be considered specialists.

Types of Test Equipment

Residential meters, meters up to 1 in. (25 mm) in size, can be tested on site in several ways to determine accuracy. The test equipment (Figure 5-10) and methods for determining the accuracy of these meters are not applicable to testing larger meters. These larger meters require specialized test equipment that can handle a wide range of flow rates and provide accurate, valid data. This equipment can either be purchased as a manufactured assembly or fabricated by the water utility. In service lines 1 in. and smaller, downstream test ports and bypass lines may not be practical; in such cases, the meter under test may be pulled from the service line and replaced by an "idler" that will provide flow to the end-user while also providing flow to a portable test rig that incorporates both the meter under test and a reference meter.

The equipment for the larger meters (Figure 5-11) is available as portable test packages and can be installed on sleds, dollies, trailers, or mounted in a van (Figures 5-12 and 5-13) or pickup truck.

Regardless of the style, these testers all contain certain basic elements that are required to properly test in-service meters. Because of the wide flow ranges involved in testing larger meters, a tester includes at least two and sometimes three meters of varying capacities. A shutoff valve is located downstream of each meter to control the flow rate for the various tests. A pressure gauge is required to check both the line pressure and the residual pressure at the tester. Sometimes resettable registers for totalizing and/or flow rates are included to reduce the time required to complete a test.

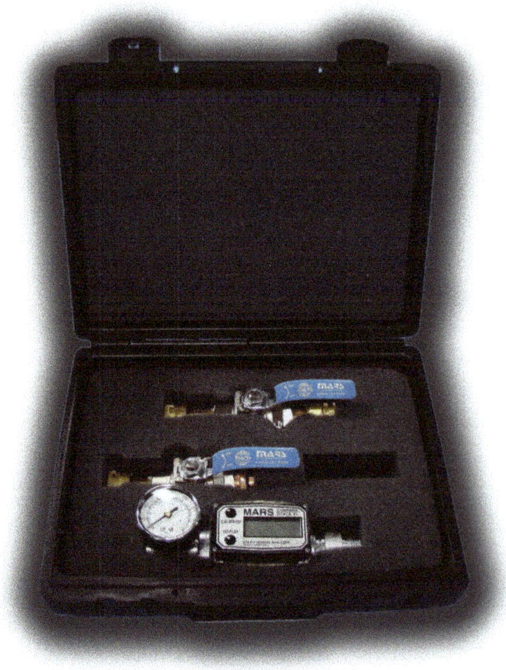

Figure 5-10 Field-test apparatus for small meters

Figure 5-11 Field testing flanged meters

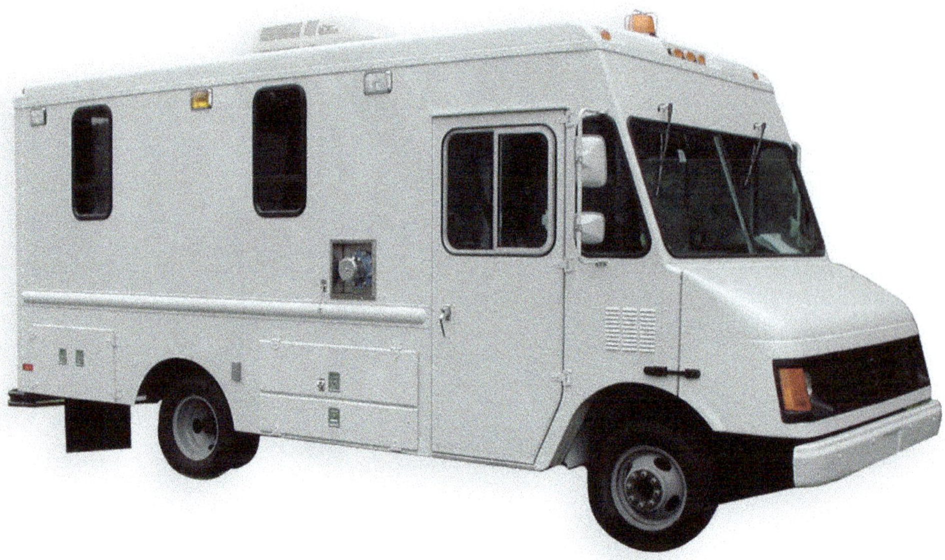

The mobile meter test van is capable of testing meter sizes 3 in. (80 mm) to 12 in. (300 mm) without removing the meters from service. The unit is self-contained, computer driven, and fully automatic.

Figure 5-12 Mobile meter test van

A portable testing system installed in a trailer.

Source: MARS Company

Figure 5-13 Mobile test unit

Flexible hoses are required to connect the test equipment to the test connection of the meter being tested. Because of the pressures and hydraulic forces, the hoses must be in good condition and positioned as straight as possible between the meter being tested and the test equipment. For larger testers, it is important that the tester itself be anchored to a vehicle or a hold-down device because at higher flow rates and higher line pressures, the tester will want to move, particularly when the flow is shut off at the end of each flow test. During a site survey to determine if a field test is practical, routing paths for the flexible hoses should be considered since hose connection leaks or ruptures could cause considerable damage. If the meter under test is located indoors, water exposure risks relative to nearby records retention, computer services, chemical storage, electrical breaker boxes, and controls should be reviewed.

The reference meters used on the testers must be capped, protected, and handled with care when not in use. They should also be tested for accuracy (and if necessary, recalibrated) periodically using standards that are traceable to a national standard such as the NIST or Measurement Canada.

Test Procedure

Before testing, it is necessary to know what the typical accuracy curve of the meter is for each specific brand, model, and size of meter being tested. This information can be obtained from the meter manufacturer's literature. A chart can be prepared that lists the flow rates (as well as test quantities) at which each meter should be tested to properly assess its operational condition. Such a chart can ensure consistent test practices. In some instances, Table 5-3 provides default test parameters in the absence of factory data. A utility can certainly test at other flow rates, such as those that may be site specific or related to manufacturers' claims.

For positive displacement meters and several other meter types, Table 5-3 of this manual provides the three flow rates (low, mid, and high) that are likely to apply to all meter brands. For turbine and propeller meters, this manual must be supplemented with the meter manufacturer's literature. For compound meters, it is important to know where the "crossover" span or range in flow rates is located so that it can be specifically tested. Similarly, the size and type of bypass meter will determine the test flow rates to use below the crossover. For this information, the manufacturer's accuracy curves are preferable. (Different brands of compounds have various crossover flow rates and different sizes/ types of bypass meters). A discussion of these crossover flow rates is beyond the scope of this manual, but recognition of crossover behavior is required when compound meters are tested.

The various suppliers of large meter test equipment provide detailed procedures for conducting accuracy tests. In general, the tester is connected to the test port of the meter setting and the line slowly and carefully flushed. The flow rate should be set to flush the testing apparatus, and the leak test should be conducted by observing the meter flow indicator as well as visually checking all fittings, connections, and isolation valves for leaks. See flush/air purge and leak concerns discussed earlier in this chapter for similar considerations in the operation of test benches.

Considerations When Testing

When testing a meter on site, the indicated throughput value of the in-service meter is compared to that of a calibrated reference meter. The calibrated reference meter has its own performance characteristics which must be reverified on a regular basis. Reference meters, like the meters they test, are not 100-percent accurate across their entire flow range, unless electronically corrected to level the accuracy curve. Even after compensating for reference meter inaccuracies, cumulative uncertainties for in-situ testing may be on the order of 1 or 2 percent over normal test flow rates and even larger at minimum test flows, unless extremely large test quantities are employed. The test objectives (see Table 5-1) will inform decisions about acceptable test uncertainties. Since meter replacement, warranty claims, and other critical decisions can be based on these tests, excessive test uncertainties should be avoided.

Once the testing begins, the testing order for mechanical meters is from the low flows to the higher flows after it has been determined that all air has been purged from the system. For electronic meters, consult the manufacturer's recommendations as the test practices may be different. Experience has shown that when most meters begin to wear or lose accuracy, it occurs at the lower flows rather than the higher. If a meter is performing accurately up through the lower 25 percent of its capacity, it will typically test accurately through the rest of the range. This is especially true on very large meters. Testing at flows greater than 25 percent of capacity of these larger meters can present problems, even in those installations where test tees or test ports can actually deliver such flows. For example, a suitable location must be found for large volumes of test water being discharged at high flow rates.

A sudden high flow rate during onsite air purge and testing of meters could reduce the supply pressure available and disturb any debris in the service line to the customer or in the meter under test. It could also reduce the water supply in a nearby portion of the system, or it could cause considerable damage if the water is not discharged properly. Standing water on trafficked surfaces can be a danger, while sheet ice on the street is a major life-threatening hazard. Great care should be taken to prevent water from freezing on the road and pedestrian surfaces near the test site. Dechlorination or other chemical treatment should be considered, as well as any relevant stormwater discharge or discharge-to-land requirements or limitations, before discharging into a storm sewer or surface body.

When setting up the tests, no test should be less than 1 min long, and the meter's sweep hand should make at least one complete revolution for mechanical registration. As noted in a Table 5-3 footnote, minimum test times of 3 min are actually recommended due to the potential for excessive test uncertainties. When time constraints do not allow for one full revolution of the sweep hand for the meter under test, note that test uncertainties/errors will increase. For example, using only one-half revolution of the sweep hand can more than double the reading error when both resolution and parallax are considered. In the case of electronic registration or an all-electronic meter, increased resolution of the indicated throughput may allow somewhat shortened test times, although this needs to be balanced against limitations due to sampling rates for all-electronic meters. Some meters incorporate a special test mode that temporarily increases the sampling rate for the meter. Users should consult the manufacturer on the operation of a test mode and the impact on test drafts and measuring uncertainties.

The residual pressure on the tester should never be less than 20 psi (140 kPa) when running a high-flow test. Also, for safety, the tester should not be operated on lines with static pressure exceeding 80 psi (550 kPa) unless provisions are made to secure the tester and discharge lines.

Detailed records are important to monitor trends in performance, along with the accuracies obtained at the various flow rates. Record the meter's registration before and after the testing, so the customer is not charged for the water used during the test. Portable computers and test software make this job easy and accurate.

The local utility must determine how inaccurate a meter can become before repairs or replacement can be justified. A cost–benefit consideration is required, based on a number of factors important to the local water system, including wastewater costs. The considerations should also include the sizing and the selection of the replacement meter. Another financial assessment should cover the updating of the meter installation/setting for future onsite testing with respect to long-term operating costs.

FACTORS CONTRIBUTING TO TEST UNCERTAINTIES

The various aspects of a meter test can each contribute to uncertainties in determining the true accuracy of the meter under test. Systemic errors may be present, but other uncertainties can be due to limitations in resolution or repeatability of either equipment or human operators. Cumulative uncertainties should be minimized wherever practical.

Under International Organization of Legal Metrology (OIML) R49 guidelines for water flow measurements, for example, test uncertainty in any individual test should be less than one-third of the allowable accuracy range for the meter being tested. With this "one-third" criterion, meter accuracy limits of 95–101.5 percent have a range in allowable accuracy of 6.5 percent, and total test uncertainty should be less than ±1.08 percent (a 2.16-percent uncertainty range versus the 6.5-percent accuracy range). For meter accuracy limits of 98.5–101.5 percent, the range in allowable accuracy is 3 percent, and test uncertainty should be less than ±0.5 percent under this same OIML criterion.

Note that adherence to this OIML criterion can be greatly affected by uncertainties associated with the meter under test itself (repeatability, resolution in reading the meter's indicated quantity of water, and so on). In testing to the relatively small test quantities given in Table 5-3, total test uncertainties have the potential to be far in excess of this OIML criterion (Sumrak et al. 2016), in large part due to uncertainties related to the meter under test. OIML-driven testing, in contrast, can have considerably smaller test uncertainty contributions from the meter under test by allowing the use of much larger test drafts with resultant individual flow test times perhaps as long as 90 min. These larger OIML test volumes are not practical for the typical water utility.

Also note that the total system uncertainty range of one-third of the allowable accuracy range is not the same as the one-fifth allowable accuracy range desirable for repeatability.

Flow Delivery System

Water must be delivered within the operating limits of the meter under test with respect to temperature, pressure, and flow rate. There must not be entrained air or other gases in the flow stream. (Mechanical meters measure the air/gas as additional water flow, while electronic meters may have errors in measurement or may not measure at all.) The actual test flow rate should closely match the selected test rate, particularly at the minimum rate. Ideally, the minimum test rate should be within 5 percent of the targeted value, although this can be quite difficult to attain, particularly during field testing. At other test rates, the actual flow during testing should be within 10 percent of the targeted value, particularly if repeatability testing is being done.

Ideally, flow should be kept constant during the course of any individual test run. Low-frequency variations in flow rate can cause some inferential flow meters, for example, to over-register. Higher-frequency changes in flow rate, pressure pulsations, or mechanical vibration can cause many types of mechanical meters to over-register as well, particularly at the minimum test rate. For some piping geometries, flow-induced vibration can also be a concern. In some cases, test benches are plumbed with multiple meters in series to increase test productivity. In such instances, the operation of one meter may create small fluctuations in flow rate or in pressure, affecting the operation of other meters in the bench. For example, oscillating piston meters may affect the performance of multijet meters concurrently placed in the same test bench. As a result, it is preferable that only like-type meters be tested together.

Electromagnetic and ultrasonic type electronic meters are inferential in nature, measuring water velocities and inferring volumetric flows based upon these velocities. (Many mechanical flow meters are also inferential in nature.) As inferential meters, they can be susceptible to flow profile disturbances and to "noisy" flow delivery systems (pressure pulsations, etc.). Care should be taken to ensure that utility test stands do not introduce such disturbances or noise at levels that are not representative of those seen in actual field installations.

Pump noise, especially that from pumps controlled by VFDs, can be particularly problematic in testing velocity meters. These VFDs can generate high-frequency pulsations in water flow. These pulses can impact meter measurement accuracy. It should be noted that the proximity of a pump for a test bench is not necessarily representative of real installation applications. Mitigation techniques can include using high volume pressurized damping tanks, using constant head elevated supply tanks for some low flows, or replacing the VFD pump drives with direct drive pumps and variable bypass circuits.

Leaks in the Test System

Water leaks occurring during a test are generally unacceptable and need to be immediately repaired. The piping layout for the test system should provide sufficient access to allow for visual checks for leaks. Errors due to leaks are generally equal to the ratio of the leak rate over the test rate. As such, small leaks can have minimal impact during maximum rate testing but generate large errors at the minimum test rate. For example, a leak exhibiting only two drips per second could result in roughly a 0.5-percent test error during a 0.25 gpm test. (A leak at two drips per second is roughly 0.0015 gpm and so represents roughly 0.6 percent of this test flow of 0.25 gpm.)

Leaks that occur upstream of the meter under test normally do not lead to test errors. An exception would be in the case of manifold systems, where multiple parallel test lines are supplied by the same source line and are eventually plumbed into the same tanks or reference meters. If an isolation valve upstream of the meter being tested in such a manifold is leaking and this leak is eventually fed back into the test line downstream of the meter, the meter under test will appear to under-register by the amount of this leak.

Leaks that occur downstream of the meter under test (or downstream of any individual meter in a test bench plumbed for multiple meters) will always lead to test errors, with the meter normally appearing to over-register by the amount of the leak. This includes leaks in drain lines for any reference tanks.

Flow Profile Disturbances

Other than positive displacement units or units with substantial internal flow-conditioning features, meters may be susceptible to changes in the distribution of water velocities across their inlet bores. To a lesser extent, meter performance may also be affected by variations in flow profiles across their outlet bores. Manufacturers provide installation guidelines in order to minimize flow profile disturbances. Ideally, flow profiles will be fully established (the same as if an infinite length of straight, properly sized pipe was at the meter inlet) and free of swirl (a corkscrew-like twisting of the flow stream, spiraling around the pipe axis). Testers should be aware of any differences in flow profiles when comparing test bench conditions with those in the actual field installation.

Misalignment at the meter end connections should be avoided. Alignment problems include a mismatch in bore sizes or gaskets with undersized bores, including gaskets that are overcompressed and so extrude inward into the meter bore. Misalignment can also be in the form of a lateral or vertical offset between the axis of the meter bore and that of the piping bore.

Changes in flow direction, changes in the flow cross-sectional area, and other disturbances can lead to skewed velocity profiles that persist for downstream lengths equal to 5, 10, or even greater multiples of the pipe diameter. When testing multiple meters in series within a single test line, upstream meters can generate disturbances that may affect performance of downstream meters. Two elbows that are not both in the same plane can lead to swirl, and this can persist for much greater lengths downstream of the second elbow.

Calibration and "Zeroing" of Test References (Tanks, Scales or Reference Meters)

Volumetric tanks or gravimetric scales can be calibrated against reference standards that are in turn traceable to national weights-and-measures standards. For example, see NIST Handbook 44 for gravimetric weighing systems and their calibration procedures. Meters used as references in meter testing would normally be calibrated in a facility using volumetric tanks or gravimetric scales and are therefore one step further removed from traceable standards. As a result, volumetric and gravimetric testing normally have the potential for smaller calibration errors than do reference meters.

While resetting (taring-out) a gravimetric system in order to start a tank calibration or a meter test is relatively easy, operation of a volumetric tank is somewhat more complicated. Volumetric tanks are normally calibrated in the initial just-drained condition, with the tank interior still wet. This mimics the conditions that would be seen in a test stand used for multiple sequential tests. So any test begun with the tank completely dry will cause a test error (the meter under test will appear to over-register). Conversely, a tank that is not fully drained before the start of a test will cause an apparent under-registration of the meter under test.

Whether a volumetric or gravimetric measure is used for a tank receiving water from the meter under test, there are similar "zeroing-out" considerations for the fill line that feeds such tanks. For every volume of water that passes through the meter under test, there should be a simultaneous equal volume of water entering the receiving tank. Any line that is not already full of water at the start of a test may result in some water never making its way to the receiving tank at test's end. Similarly, any line that very slowly drains into the receiving tank could be contributing pretest remnant water to the receiving tank or still be slowly adding test water to the tank after the final read on the tank has been made. Lines that fill the receiving tanks normally end in a gooseneck or other feature to ensure full lines between tests and to minimize any drain-into-tank times at the end of the test.

Resolution Limitations in Reading Meters

All devices have limits on the resolution at which they can be read. In a meter accuracy test, there will be uncertainties related to reading the meter under test, as well as uncertainties in reading the reference device. These uncertainties can be the major contributors to total test uncertainty, even if the test equipment is calibrated and operated properly.

Reading a mechanical register on the meter under test (or on a reference meter) can have several sources of error, including parallax (if the reader's eye is not in the vertical plane that passes through the test index sweep hand) and "jump" in the sweep hand movement if gear train backlash has not been "taken up." But even with proper procedures, there are limitations in the repeatability of reading the test hand position with respect to the printed index circle. For example, a ⅝-in. meter may have a center-sweep, straight-reading register with a 10-gal test circle. The test circle would have 100 graduations. In a minimum rate test using a 10-gal test quantity, the sweep hand might incrementally move past something like 95 or 100 or 103 of these graduations. If the reader records sweep hand movement only to the nearest whole number of increments, there may be a reading error of roughly ½ increment (or more) of the test index (representing roughly a ±0.5-percent potential uncertainty versus the size of the test quantity) at the beginning of the test and a similar error at the end of the test. If the reader interpolates between the printed graduations on the test index, overall reading uncertainty (even after including slight parallax, general visual acuity, and other parameters) might be reduced to something on the order of ±0.3 percent.

There are some mechanical registers/encoders that have smaller, higher-resolution test indices. If the physical spacing between graduations is small, interpolating between these graduations may not be practical (or needed). For electronic registration, the increments by which the totalizer is advanced are smaller than the increments on a center-sweep, straight-reading mechanical register.

Note that if a shortened test is attempted, resulting in the test index only incrementing by one-half revolution per 50 graduations (for example), reading uncertainties as a percentage of total test volume would double (or more than double, if parallax issues also worsen). Conversely, for larger test drafts that result in multiple revolutions of the test index, uncertainties in reading a register become progressively smaller (as a percentage of total test volume).

Reading issues may be further complicated if register test indices are interrupted or partially obstructed. In some cases, add-on test rings can be placed on the register to supplement or substitute for the register test index. Alternatively, it may be possible to "run-up" the register sweep hand and select test draft/volume sizes so that test start and test end are both in locations where the index is usable.

These same meter reading uncertainties would also apply to reading a reference meter if it has a mechanical register.

Resolution Limitations, Volumetric or Gravimetric References

Reading the sight glass on a volumetric tank can also have several sources of error, including failure to properly read to the bottom of the meniscus, a dirty/scratched tube that deforms the meniscus, or parallax errors if the reader's eye is not in the same horizontal plane as the meniscus. But even with proper procedures, there are limitations in the ability to read the water level in the sight glass, particularly for straight cylindrical tanks. For a tank that is at its full calibrated volume when the water level is at 4 ft (1.2 m), every ½-in. (13-mm) change in level represents a 1 percent difference in volume. If the level scale graduations around this calibration point are at 1 percent intervals, accurate reading of the tank volume will almost always involve interpolating between the calibration point and the actual level of the meniscus. If this interpolation can be done within ⅛ in. (3 mm), this represents a ±0.25-percent uncertainty. Further uncertainty results from the unknown degree of accuracy between the calibration point on the tank scale and the other, noncalibrated graduations on this same scale. If a shortened test time is attempted through the use of only a 50-percent tank volume, the reading resolution uncertainty at the end of the test would double (as a percentage of total test volume), but of even greater concern is the fact that many tanks are not independently calibrated at this 50-percent mark.

Water level reading resolution can be greatly improved if the tank is necked-down at the calibration level (this geometry is often called an "onion" tank). For example, if the 4-ft-high cylindrical tank discussed above is replaced by a tank that is at twice the original diameter for most of its length but then half the original diameter as the tank height approaches the calibration point, overall tank height is greatly reduced while reading resolution at the calibration point would be four times better than in the original tank. Such tanks are often used as transfer standards when calibrating straight cylindrical tanks but are less common as the actual reference tanks in a water meter test bench.

Gravimetric tanks normally have a direct digital readout for the weight of the water in the tank. Reading resolution is typically better than for volumetric sight glasses. However, there must then be an accurate calculation of water density for the specific test conditions to compare the volumetric reading of the meter under test with the gravimetric reading of the water that has passed into the tank. Uncertainly in determining the water density can be as great as 0.05–0.1 percent, depending upon the methods used.

Meter Sampling Rate

Meter sampling rate applies only to electronic meters. Similar to reading resolution, the sampling rate of the meter relative to the test duration will create uncertainty with regard to repeatability. As an example, if a meter has a sampling rate of once per second, and the duration of the test is only 180 s, there is a 1-s uncertainty at the start and stop of the flow test. This results in an uncertainty of ±2/180 or ±1.1 percent.

Static Versus Dynamic Testing

Meter testing is normally done in a static manner—i.e., the test starts with zero flow (and empty reference tanks). The targeted test flow is then quickly established and maintained for most of the test and then flow is smoothly brought down to zero as the targeted test volume is approached. This process is termed "static" because the readings at the start and end of the test are taken during no-flow conditions. There are additional test errors that may be introduced during static testing. For example, significant test time may be at flows other than the actual targeted rate. Also, too abrupt a stop in flow at the end of the test can reverse the take-up in mechanical gear-train backlash that preceded the start of the test, so that the meter under test will indicate some additional registration. And there is the potential for some inferential-type meters to coast momentarily after an abrupt flow stoppage, adding to their apparent flow registration. All of these potential errors are exacerbated by shorter test times.

The alternative method of dynamic testing is not available or viable for most meter shops. However, utilities may be exposed to this approach in dealings with national metrology labs, third-party test facilities, or manufacturers' test lines. The basic test premise involves recording meter registration and output from the test references "on the fly," while the targeted flow rate is being maintained. This relies on a significant degree of control automation and specialized equipment perhaps dedicated to specific meter designs.

Cumulative Effects from Errors and Uncertainties

For the same individual meter, test results from multiple test facilities are expected to vary somewhat. This can be attributed to the inherent repeatability of the meter, minor and unavoidable errors in reference calibrations, and reading resolution limitations for both the meter and test references. However, variations can increase greatly if there are issues with the calibrations, test practices, or test facility conditions. In such cases, the accuracy test results may be invalid.

For example, there was an extensive study using the Utah Water Research Laboratory (UWRL) and 12 utility test facilities (Shields et al. 2013). This involved sets of 10 residential meters (of various metering technologies) first tested at UWRL, then tested at a given utility's facility, and finally retested at UWRL. While the before and after results at UWRL were quite consistent, the test results at the utility benches were significantly variable, even after discounting the results from the sole utility that used a portable test rig to test meters individually and only assessing results from the remaining 11 utility test benches in the study. From the specific test results that are summarized below, it can be seen that fairly large test discrepancies occurred in some utility test bench tests, reinforcing the need to periodically assess and then reduce these sources of test errors.

In just reporting the 10 meter averages, the authors' data involving the 11 remaining utility test benches included

- minimum flow: only four utilities had results within roughly 1.5 percent of the UWRL;
- intermediate flow: four utilities had results within roughly 0.5 percent of those at the UWRL, and the remaining utilities were within roughly 1 percent; and
- maximum flow: seven utilities were within roughly 0.25 percent of the UWRL results, and another two utilities were within roughly 0.5 percent.

When assessing the absolute values of discrepancies between UWRL results and those for the utility benches for each individual meter and then averaging these absolute values, the discrepancies were (understandably) much larger. For example, if a utility result was 1 percent higher than UWRL for one meter and 1 percent lower for another meter at a given flow, the discrepancy for the average of these two meters would be zero, but the average of the individual absolute discrepancies would be 1 percent. While discrepancies in the raw average for 10 meters might indicate errors in calibration, leaks, and other systematic problems, the average of the absolute values in discrepancies may, in addition, indicate random issues that may be associated with repeatability problems.

From these averages of "absolute discrepancy," the assessment yielded the following data:

- for minimum flow, only one utility was within 1.5 percent of UWRL results;
- for intermediate flow, only one utility was within 0.5 percent of UWRL results;
- for maximum flow: four utilities were within roughly 0.5 percent of the UWRL results; and
- an additional four utilities were within 0.5–1.0 percent.

The initial assessment of uncertainties in test equipment and practices can be repeatability testing within a single test bench. This typically involves performing three sequential tests at the same flow rate for the same meter (or the same string of multiple meters in a larger test bench). The results should certainly all fall within a range that is less than one-third of the total range in accuracy allowed for the meter under test. Note that for some electronic meters, repeatability testing may indicate large test uncertainties unless increased test volumes and/or high-resolution "test-only" modes are used.

Repeatability testing, however, will not be able to check for systematic, consistent errors in test equipment or practices: e.g., chronic leaks, flow profile disturbances, or errors in tank/scale calibrations. In a test of multiple meters simultaneously in a single bench, if all meters are calculated to have similar unexpected accuracy errors, a systematic testing error may be indicated. The presence of such errors might also be indicated but not diagnosed by testing the same meters in multiple test stands (i.e., "round robin testing" per Shields et al. 2013). Ultimately, one may need to assess each individual aspect of a meter test to minimize the various sources of test uncertainties.

REFERENCES

Shields, D.J., S.L. Barfuss, M.C. Johnson, and C.C. Hannah. 2013. "Meter-testing Methods Matter." *Journal AWWA*, 105:11:E664–E670.

Sumrak, M.L., M.C. Johnson, and S.L. Barfuss. 2016. "Analysis of Meter Registry Uncertainty" *Journal AWWA*, 108:9:E467–E473.

Tao, P. 1982. "Statistical Sampling Technique for Controlling the Accuracy of Small Water Meters." *Journal AWWA*, 74:6:296–304.

Chapter 6

Maintenance and Rebuilding/Repairing of Water Meters

INTRODUCTION

There are millions of water meters in service in the United States. While these meters represent a sizable investment for water utilities, their real value cannot be realized unless they are functioning efficiently. Many of the meters presently in service are more than 20 years old.

In the past, the design of these meters was complex and cumbersome, and their performance, even when new, was not comparable to modern meter designs. Modern technology is providing meters with tighter accuracy specifications, lower head loss, and greatly simplified maintenance.

Most modern meters (except electronic solid-state meters) contain magnetic couplings and sealed registers. This design concept eliminates the highly maintenance-prone stuffing box; heavy, wet-running gear train; and exposed register. Additionally, most modern meters contain synthetic polymer components that are resistant to corrosion.

The success of the synthetic polymer chamber and magnetic coupled sealed registers has reduced maintenance costs dramatically. Meter shops are still the best opportunity to maintain meter accuracy and protect the related revenue (Figure 6-1). The skills needed to maintain water meters have changed. Older complex meters with gear trains required skilled craftsman, whereas the modern sealed register meters are made of components that can be quickly exchanged by a technician with limited training. Utilities usually dispose of the outmoded meters by selling them as scrap or through a trade-in allowance against new

Courtesy: Elster AMCO Water Inc.

Figure 6-1 A meter technician maintaining modern concept water meters (left: endurance test, right: accuracy test)

meter purchases. Although a period of regular replacement of meters instead of repairs was used for a time, modern meters are more advanced, more accurate, and too costly to simply scrap on a regular basis. In addition, the need to test meters has not gone away, given customer issues and the need to ensure revenue protection.

Water meters experience many conditions during their deployment, including infrequent or low levels of usage. As such, time in service, like consumption levels, might not be the best predictors of meter accuracy but could be indicative of more intensive investigation, including testing to determine actual accuracy performance. It is rarely advisable to repair outdated or obsolete designs, but modern meters are modular in design, making them easier to repair. Adding automated reading technologies to the meter or investing in electronic meter technologies that have no repairable components makes the metering system too expensive to throw away without a proper evaluation and accuracy testing to determine serviceability. An evaluation of cost per meter repaired is valuable in making this decision.

METER REPLACEMENT

Electronic solid-state or static meters generally retain their accuracy for the life of the battery. A planned meter replacement program for mechanical meters can be implemented over a given number of years—for example, 10 percent of the meters each year over 10 years or 20 percent per year over 5 years—so that all replaced meters in the system will be of the more efficient, modern design. Concurrently, all additional meters purchased for expansion will be of the modern design, and the entire system will be upgraded on a planned schedule. This provides the utility with two advantages:

- the meters in service are far more efficient in their ability to measure all water consumed and to recover maximum revenue for the utility and

- the expenses of maintaining, stocking, staffing, and operating a traditional repair shop are greatly reduced by a regular meter maintenance program.

Modern magnetic-drive meters and electronic meters contain very few service parts; therefore, the ongoing maintenance becomes one of simply replacing the major components rather than laboriously hand reworking, cleaning, and reassembling a multitude of miscellaneous parts. All major mechanical meter manufacturers offer factory-built, measuring-assembly replacements that can be installed by an easily trained worker. Most also offer replacement register assemblies. The materials used in the measuring chambers and registers have eliminated cleaning or replacement due to corrosion.

The traditional meter shop has and will continue to change to match the modern technology used to create today's water meter. There will always be a requirement for maintenance and upgrading of existing meters. It will be done on a very simplified replacement-by-assembly basis in the near term and possibly by adopting the replacement-meter concept in the long term if inflation and increased labor costs so dictate.

METER MAINTENANCE AND REPAIR

Detailing of specific procedures is impossible in a text of this scope because of the variations in design of the different brands. On request, established manufacturers will provide specialized information if required on repairing meters of their own make. They should be consulted in advance for procedures especially applicable to their meters.

Dismantling and Cleaning

It is recommended shop practice to clean the exterior of the meters first to avoid introduction of dirt into the meter repair shop. Sand- or bead-blasting is an efficient cleaning method. Wire brushing and hand scrubbing can also be used. Sandblasting may be used only on outer casings.

DISASSEMBLING

After the exterior cleaning is completed, the meter should be dismantled into its basic components of the shell or case, usually an upper and a lower part, the register, and the measuring chamber.

Power tools with torque-limiting capability are useful for speeding up assembly and disassembly of meters. The use of pneumatic rather than electric tools is recommended for safety.

Casings

Bronze casings should be checked for distortion caused by frost. If damaged, the casing must be replaced. The meter should then be cleaned of all gasketing material with a suitable tool.

The threads on the inlet and outlet spuds should then be checked. If they are nicked or otherwise damaged, they can be repaired by running a rethreading die over them. Note that the threads on all meters having external threads are straight threads, rather than tapered, and the thread dimensions must be as required in ANSI/AWWA C700 Cold-Water Meters—Displacement Type, Metal Alloy Main Case; C708 Cold-Water Meters—Multijet Type; and C715 Cold-Water Meters—Electromagnetic and Ultrasonic Type for Revenue Applications, which conform to ASME B1.20.1.

Register Boxes

Modern meter registers that are damaged should be replaced with a new sealed register assembly. If the lid or serial number plate is bent or broken at the hinge, it should be replaced.

Modern water meters have fully sealed gears and cannot be serviced. Older gear-trained-equipped meters are not upgradable to modern sealed registers or automatic meter reading (AMR) equipment and are expensive parts. These units are also obsolete and should be replaced. The scrap brass should be sold to recover part of the original investment.

Measuring Chambers

The measuring chamber is the heart of the meter, and it is this chamber that determines the meter's characteristics and accuracy (Figures 6-2 and 6-3). Measuring chambers are of several types, commonly referred to as the nutating-disc type, oscillating-piston, multijet, turbine, fluidic-oscillator, singlejet, and others used in large meters.

Several meter manufacturers have available chamber-exchange programs with guaranteed accuracy for moderate costs.

Reassembly and Testing

Reassembly involves the assembly of components, including the main case, measuring chamber, and register assembly. Bottom plates can be polymer, cast iron, or bronze. Frost-proof bottom plates are designed to break and protect the higher value meter main case. The measuring chamber may incur some damage and require replacement.

MECHANICAL-DRIVE METER REASSEMBLY

Some utilities still use some mechanical drive meters, but these are becoming obsolete and are being replaced as new designs are introduced.

Every repaired meter should be tested for accuracy. Recommended accuracy standards for various rates of flow in repaired meters are given in Chapter 5.

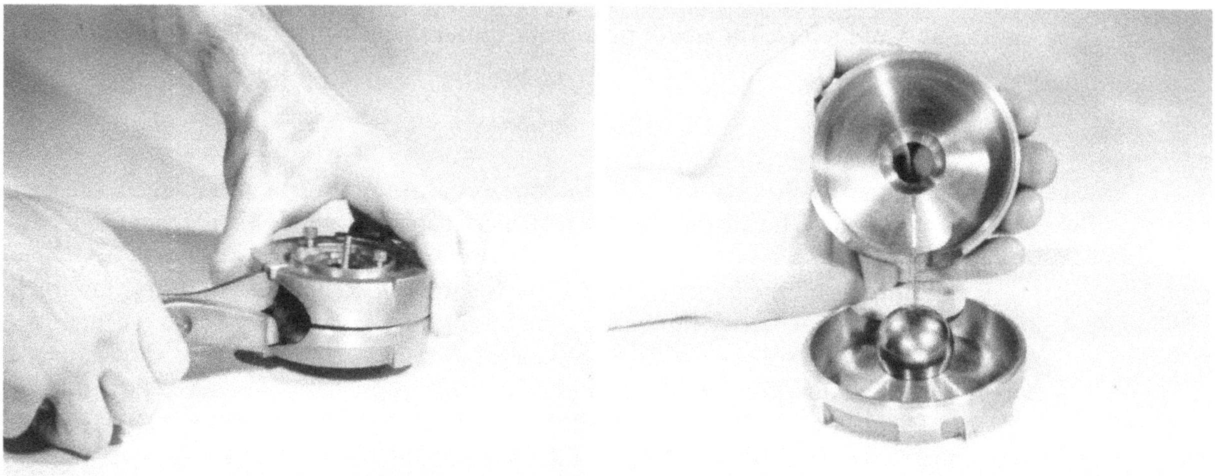

A special prying tool (left) is used to open a two-part chamber. A metal ball gauge (right) may be used to check the socket in the measuring chamber for wear. This is an early predecessor to the modern polymer disc chamber.

Figure 6-2 Opening of disc chamber and gauging of chamber ball socket

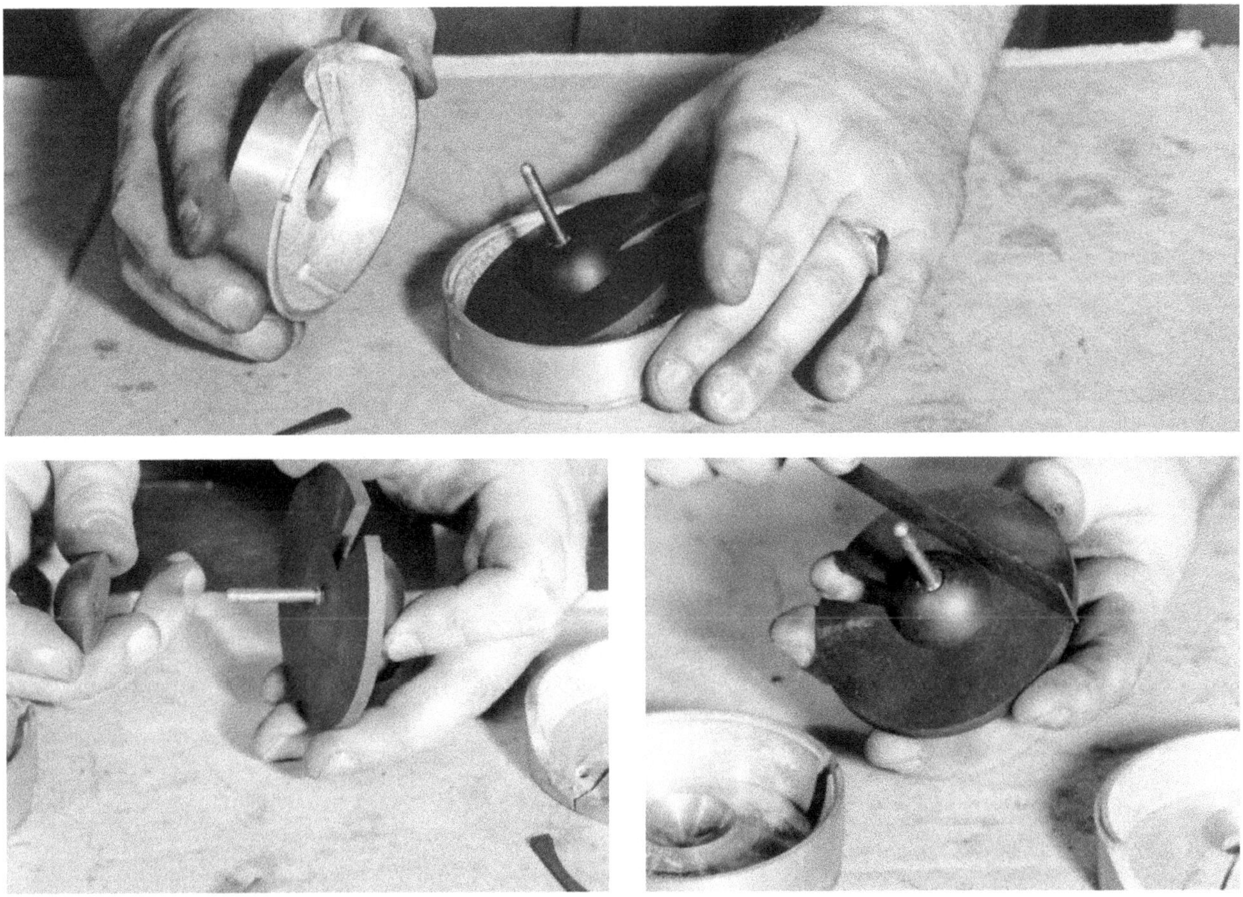

Correct fit and clearance in the disc and chamber largely determine the accuracy of the meter.

Figure 6-3 Examination and assembly of chamber and disc

REPAIR OF MAGNETIC-DRIVE METERS

Registers

There are two types of registers used on magnetic-drive meters. One is a permanently sealed unit that is not repairable. If damaged, this entire unit must be returned to the manufacturer and repaired or replaced. The majority of modern meters are produced with permanently sealed registers. In the repairable register, the register housing is fastened to the main casing with screws and sealed with an O-ring.

CONCLUSION

Due to modern meter engineering design, material costs, and manufacturers' programs, many meter repair shops have limited their scope of repairs. Meter testing for revenue protection and minor repairs are the new state of the art. Each water utility must study the economics of meter repair versus meter replacement. Modern meters are produced with very few serviceable modular parts; therefore, replacement programs are more feasible than repair programs. Although cost-effective repairs may be limited in scope, the maintenance through testing of water meters should continue to ensure equity among customers and to ensure minimal revenue loss resulting from under-registration of meters. Modern meters with AMR or advanced metering infrastructure (AMI) systems are replacing many older, obsolete designs. These meters are much more accurate and can produce more revenue due to their accuracy and ease of service. Even utilities that are not yet using modern AMR/AMI systems should consider meters that are easily convertible to these systems in the future.

Chapter 7

Meter Shop/Lab Layout, Equipment, and Management

INTRODUCTION

Water has become more widely recognized as an important natural resource. As such, water utilities need to develop or expand existing effective meter repair and replacement programs to meet water resource management challenges. It is important to develop an efficient meter repair shop through proper equipment selection and meter shop layout along with updated data management and analytical techniques. Many meter shops are housed in existing facilities, and space constraints place operational restrictions on management. When the opportunity arises to modernize a facility through remodeling or rebuilding, management should be prepared to optimize meter departments with the proper equipment and an efficient layout. Architectural consultants can provide valuable assistance, but they must be advised on the necessary features and functions of a meter shop to provide for an adequate supply of test water. In cases with substantial test volume throughput or with water effluent concerns, use of recirculation systems should also be considered. Also, federal, state or provincial, territorial, and local agencies may regulate the discharge of water (and waste) to streams, waterways, and land. For example, the US Environmental Protection Agency (USEPA) regulates discharges into streams and waterways of the United States through its National Pollutant Discharge Elimination System. All meter testing programs should ensure compliance with any relevant discharge regulations.

Meter shops come in all sizes and shapes with the capability of handling from a few dozen to many thousands of meters each year. Adequate space is essential to provide for an efficient operation. Conditions that can affect the size, equipment, and layout of a meter repair shop include the following.

- Total number of water meters in the system and the projection for future growth
- Ability to test, repair, or replace modern electronics like automatic meter reading (AMR)/ advanced metering infrastructure (AMI)
- Variety of makes, sizes, and types of water meters to be serviced
- Extent of repair capabilities desired and/or required
- Knowledge to recommend repairing, modernizing, or salvaging aging meter models
- Water quality in the system, as this can have an effect on meter corrosion, lining dezincification, and wear
- Local practice or the state requirements specifying how often meters must be tested and local code requirements for properly approved installations
- Private ownership of meters, which may call for the identification and the segregation of meters to ensure that they can be returned to the proper owners
- Requirements for private owners to test or repair aging meters using authorized or certified installers
- Requirements that all privately owned meters can be tested at any time by the water utility to ensure correct revenue maintenance
- Special cleaning/painting requirements to satisfy USEPA requirements and company policy
- Ventilation system to exhaust fumes from meter shops and enclosed spaces
- General plant services available, including
  - air at appropriate pressure and volume sufficient to handle all equipment
  - water at appropriate pressure and volume for high flow rate test requirements (Consider a dedicated supply line specifically for the test bench to reduce pressure fluctuations.)
  - electric capacity and lighting available for equipment and workstations
  - water drains and/or recycling capabilities
  - internet connectivity for system maintenance, troubleshooting, and updates (These can include Ethernet connections [wired technology] as well as Wi-Fi and Bluetooth connectivity [wireless technologies].)
  - accessibility to the meter shop for equipment installation, removal, and maintenance as well as ensuring ease of transporting meters to and from the meter shop
- Meter inventory handling and control procedures

METER SHOP EQUIPMENT

Because reliable equipment is essential for producing quality work, basic shop equipment should include testing equipment, cleaning equipment, and workstations.

Testing Equipment

The following sections provide a discussion of a variety of equipment used in testing water-meter performance.

Test bench for meter sizes ⅝ in. (15 mm) through 2 in. (50 mm). The size of the test bench and the number of meters that can be tested at one time are determined by the workload. A large meter shop may need one or more test benches for the most common size of residential and larger commercial/industrial meters. Factory-built test equipment for residential size meters is readily available.

Each test bench should be equipped with appropriate valves, pressure gauges, and a rate-of-flow measuring device. Hydraulic clamping to secure meters in the test bench and automatic water shutoff at preset testing levels will increase efficiency.

Test uncertainties can be reduced through the use of gravimetric tanks. Weighting system scales should be calibrated according to manufacturers' recommendations to ensure accuracy and comply with applicable national standards.

Test station for meter sizes 3 in. (80 mm) and larger. Equipment for testing 3 in. (80 mm) and larger meters varies greatly based on the quantity of various size and type of water meters. Many test benches can safely test meters up to 30 in. in diameter. Test stands can be designed and fabricated locally but should always include provisions for convenient exchange of meters with different dimensions. Overhead lifting devices are usually needed in addition to flow measuring devices, pressure gauges, and control valves.

Test tanks. Calibrated tanks of various sizes should be used together with the test bench to contain the water used for testing. In addition, sight tubes on the tanks are necessary to gauge the water level on a volume basis. Gravimetric systems can be used with the tanks for a higher level of accuracy and should comply with applicable national standards. Regular calibration by a qualified third party of the volumetric and gravimetric systems is highly recommended to ensure accuracy.

Safety Issues

Testing operations can be dangerous and require technician training to ensure maximum safety and production throughput. Water moving at high velocity through a testing device can impart destructive water hammer shocks unless precautions are taken to eliminate these high shocks to a level that will not destroy the meters, the testing equipment, or the water mains in the vicinity of the meter shop.

Most testing operations are relatively safe because the water velocity is controlled by the test bench or the meters have restrictions.

Larger test systems operated by trained technicians are also safe when these simple rules are followed.

- No test system can be operated at working pressures greater than 150 psi, including any water hammer or pressure surge, unless permitted by a specific standard. ANSI/AWWA C703 Cold-Water Meters—Fire-Service Type should be referenced for the exception. Pressure-reducing devices must be used ahead of the first appliance in the system, including the test equipment and control valves.
- Quick-closing valves, such as lever-operated ball or butterfly valves, cannot be used without motor operations calculated at a closing rate that will eliminate water hammer.
- Meters with case damage cannot be tested safely under pressure because of the possible failure of the case causing projectiles, high-pressure water damage, or injury.

- Emergency shutoffs should be designed to safely shut down all valves and pumps.
- Lifting equipment and other special meter-handling devices prevent lifting injuries and other injuries caused by falling meters. It is important to design the shop and the meter-handling equipment to allow technicians to handle and test all meter sizes safely.
- Field testing has its own special safety issues because of such factors as traffic and pedestrian volume and space confinement. Adequate lifting equipment and service vehicles equipped with safety harnesses and lifting devices should be considered and required.

Meter-Cleaning Equipment

Compressed air with an abrasive or a mild biodegradable soap solution is favorable for cleaning meter parts in preparation for repair and reassembly. Glass beads are the most practical abrasive. Adjusting the air pressure according to the severity of the cleaning problem eliminates possible damage to machined bronze surfaces.

Honing should not be used on plastic or delicate materials. These pieces should be cleaned by other methods described below.

In areas where waters are mild and deposits are soft in nature, hand brushing with warm water and a detergent may be sufficient for cleaning interior meter parts.

Ultrasonic cleaning systems can be used for preparing interior meter parts for repair and reassembly.

All cleaning materials should be disposed of properly. Acid baths or washes or other methods of aggressive chemical cleaning should not be used. These types of cleaning methods are detrimental to the meter materials, violate USEPA regulations, and may cause significant harm to the user, environment, and water supply.

Painting has been used for more than 50 years in meter shops in North America. Because of the many hazards present in paint fumes, odors, and paint spray damage, most shops have eliminated all painting booths and the related expense. Water meters with a clean bead-blasted outer surface are as reliable as a meter painted with brass-colored paint.

Workstations

Workstations should be equipped with

- a workbench with a steel or hardwood working surface for each meter repair station,
- a sink with hot and cold running water at each workbench,
- tool compartments that lock,
- compressed air at appropriate pressure and volume at each bench for pneumatic tool use,
- storage bins and drawers for meter parts,
- essential tools, and
- any special equipment needed to test AMR/AMI devices.

SHOP LAYOUT

Most successful industries have conducted studies in attempts to reduce unnecessary employee movements. These studies have established generally accepted principles applicable to any field of production. Locating equipment to ensure smooth work flow with minimum distances between stations while providing adequate transportation of materials plus temporary storage are the objectives in planning shop layout.

Meter maintenance procedures may vary from shop to shop depending on control procedures (with respect to meters removed) prior to the meters entering the meter shop. In general, however, the procedure for an incoming meter is for the meter to be

- checked for reading and identified as the meter stated on the removal order and
- tested for incoming efficiency.

The reasons for meter removal are many, and these reasons usually dictate handling procedures. If the meter is removed for regular replacement due to obsolete technology—i.e., the meter is inoperative or not compatible with new AMR/AMI systems—management may require a test for revenue analysis purposes or simply place the replaced meter in a scrap holding area for future sale to recycling contractors. If the meter involves a customer complaint or a dispute, special handling and storage is required until the matter is resolved. The decision is then based on the utility's evaluated experience, the meter's registration, its test performance, and normal service life.

When repairs are appropriate, the procedure is as follows, with variations due to local requirements and applicable regulatory standards. The meter to be repaired is

1. disassembled and cleaned,
2. inspected for reassembly with non-serviceable parts being rejected,
3. fitted with needed new parts and reassembled,
4. tested for performance and accuracy of registration, and
5. sealed to guard against tampering.

Work flow is the key element to proper shop layout. Depending on the space available, a shop can be designed in a U-shape or in a straight-line fashion, with the old meters going in on one side and the rebuilt meters going out of the other. Workstations need to be separated enough so that the mechanics do not interfere with each other, but close enough to avoid unnecessary steps between work areas. Provide sufficient separation for reuse and nonpotable meters to avoid cross-contamination.

Material handling in a meter test and repair shop is one of the important facets of efficient operation. Depending on the size of the shop, meter processing can be either a continuous or batch operation. Most shops, because of their smaller size, process their meters in batches. The optimum batch is a one test-stand load. Smaller shops that test only small meters under 2 in. may have a single mechanic perform each operation required to complete the job. Larger shops that test large meters usually use a two-person team to enhance safety, production speed, and technical expertise.

Transporting meters from processing station to processing station can be conveniently done on carts. In addition to their use as material-handling devices, the carts also serve as in-process storage between stations in the repair cycle.

In general, available space and the bench location dictate shop layout. The best rule of thumb is to design the shop layout to receive meters from the field or from manufacturers through the main receiving area. From there, meters can be diverted through the test and repair section, ending up in a secure storage area awaiting reinstallation. New meters can be inspected and subjected to receiving procedures dictated by company policy and then sent to secure storage. The shop should be laid out with equipment or workstations in the order in which they are used. With the test stand at the end of the process in a U-shaped layout, it is also close to the beginning of the process to facilitate testing of incoming meters. The most frequently used replacement parts should be stored at the workstations.

Good lighting is essential for the worker to be able to do a first-class job.

Each repair bench should have the most frequently used tools on hand, as well as a small sink for special cleanup.

If all water utilities had followed the same processing cycle in repairing meters, standards for equipment and shop layout would have been established in the past. These standards would have varied because of workload, new developments in equipment, and improvements in meter design.

Three shop layouts that satisfy smooth work-flow requirements for the previously detailed meter-repair procedure are shown in Figures 7-1 through 7-3. U-shaped layouts allow the incoming meters to move smoothly through the shop. Transportation from station to station and temporary storage at any point is provided by a meter cart. Back tracking and cross movements are eliminated. Roller conveyor belts would be practical in a large shop for moving repaired meters from the repair benches to the test bench. In these figures, the center floor area has been kept open for additional equipment or additional work area if the workload were to increase or repair procedures were to change. The number of repair benches could be doubled without disturbing the pattern of work flow.

Here are a few key items to keep in mind when designing your meter shop.

- The kind of equipment that will be required: e.g., bench, water recirculation system, tanks, and other equipment
- Proper floor drains that are sized correctly to handle the water flow
- Proper water supply to the bench or benches to ensure proper test flow and pressure
- Proper electrical capacity for pumps and testing equipment
- Adequate supply of compressed air for running the equipment or meter refurbishment (Locating the compressor in an adjacent room or enclosure is recommended to keep noise levels at an acceptable level.)
- Internet connectivity (Ethernet, Wi-Fi, Bluetooth), if available, which can assist with equipment diagnostics, support, and upgrades
- Access to the meter shop as appropriate for equipment installation, removal, and maintenance
- Testing needs that may change over time and allow for adequate growth or expansion

An experienced meter repair supervisor is essential to the operations of a meter maintenance shop. Supervisors must be receptive to suggestions for work simplification from any worker. Obtaining special new equipment and tools, as availability and need are proven, will result in improved quality and increased quantity of work performed. Keeping all parts of any one meter in a metal or fiberglass pan, from disassembly to reassembly, is an example of work simplification. Scheduling repair to one make of meter during any one time period, if the work load permits, simplifies the whole repair procedure. Reviewing test data from the test benches will result in quality control and adherence to established repair standards.

Proper recordkeeping has been revolutionized by the use of computers, and the modern meter shop needs to be tied into the water department information system. Complete layout planning requires consideration and space for computer data entry, storage, and retrieval, as well as remote access for equipment diagnostics, support, and upgrades.

To summarize, each meter shop will vary in design and layout based on the facilities available and the work to be performed. After those factors are determined and the equipment is selected, a layout can be developed based on time-proven principles of efficient material flow. Considerations to recognize include new health and safety regulations and new environmental laws regarding waste disposal. Plans for new shop layout should be reviewed with management to ensure compliance.

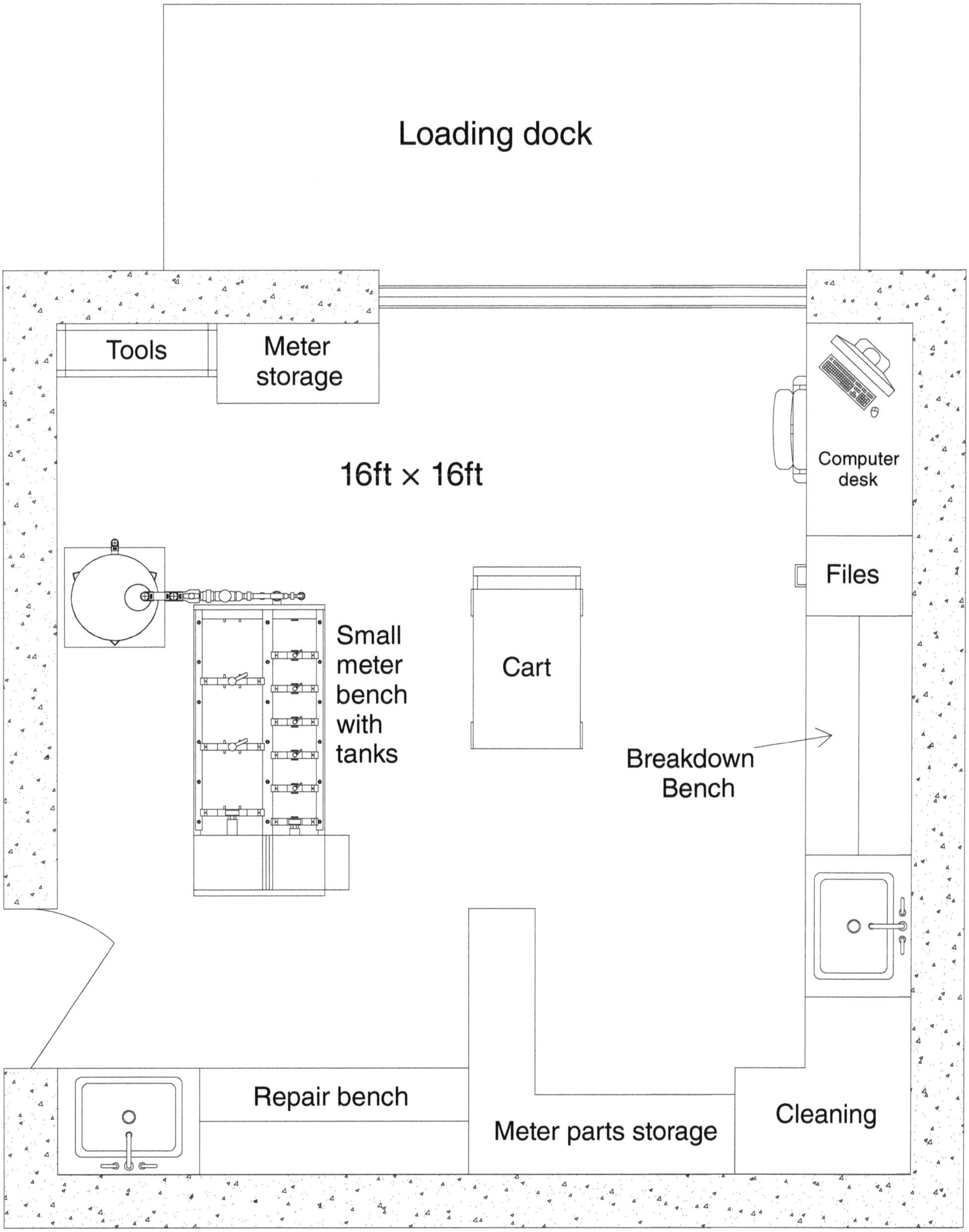

Illustration by Chris Mastic, MARS Company

Figure 7-1 Suggested layout for small modern repair shop

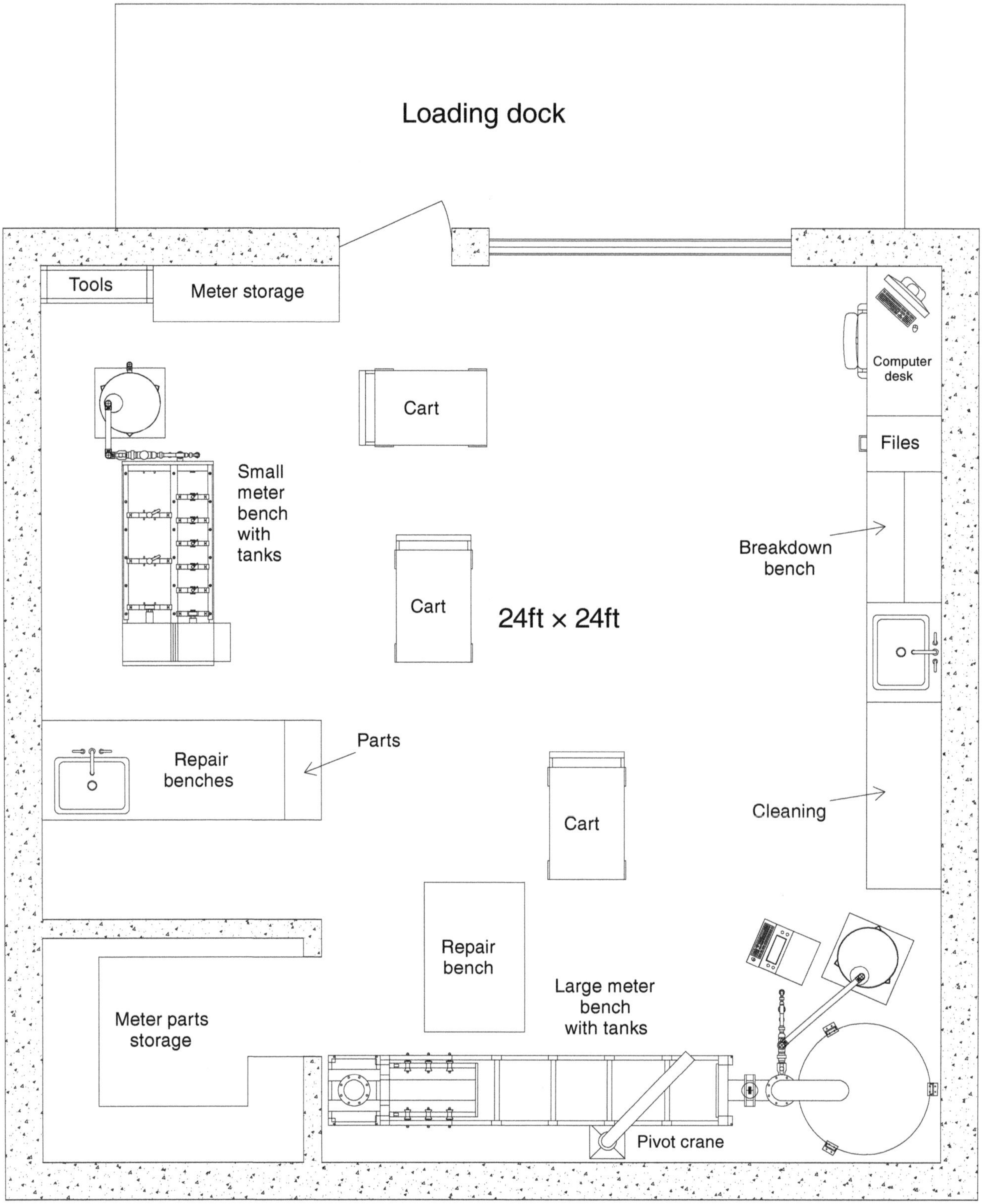

Illustration by Chris Mastic, MARS Company

Figure 7-2 Suggested layout for medium modern repair shop

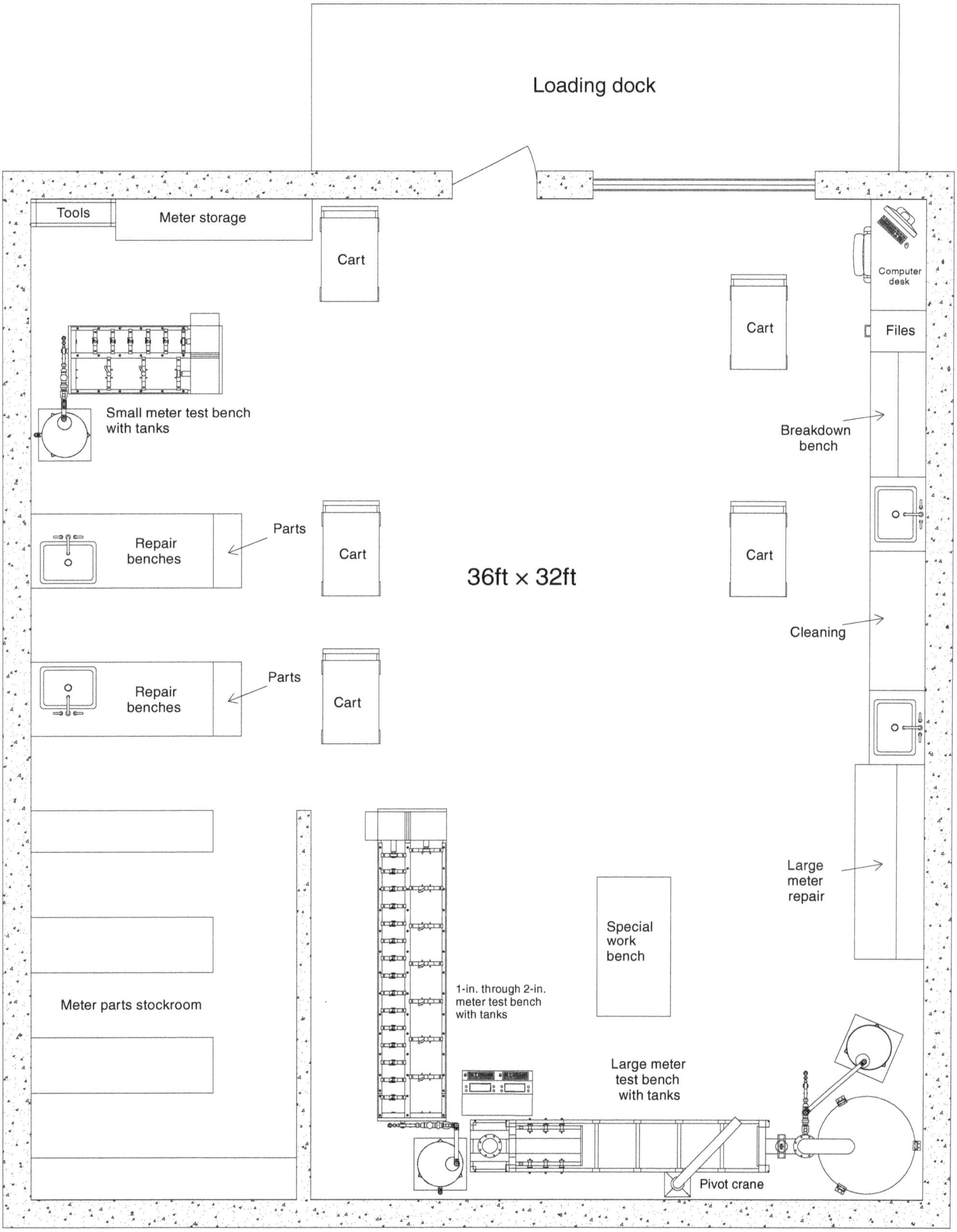

Illustration by Chris Mastic, MARS Company

Figure 7-3 Suggested layout for large modern repair shop

This page intentionally blank.

Chapter 8

Meter Records

INTRODUCTION

A suitable system of meter records provides full and complete information regarding the installation, repair, testing, and data collection for each meter. Before advancements in technology, considerable time and effort were needed to compile and maintain these records. Continual improvements in computing technology and meter management software, including the implementation of a utility's customer information system (CIS) and their procedures, have not only improved the recordkeeping and database management of meter test results, but have also provided the meter test facility the ability to track water loss (apparent and real) and analysis to make informed, economical decisions.

ELEMENTS OF THE MODERN RECORD SYSTEM

Initially, meter records should provide basic data for each meter, such as size, make, type, date of purchase, meter location, testing data, and repair data. One method of maintaining these records is through the implementation of a fully networked, computer-based tracking system. Because a percentage of the previously mentioned data is often available on a water utility's CIS, it is practical to develop a computer-based meter management system to combine customer information with meter records for a usable tool that is part of the main data system. The purchase date, testing, repair information, and previous locations can be entered into this system. In addition, connectivity between the meter test facility and the CIS can be created between the meter testing shop and the data system for use in generating reports and conducting research.

Each meter purchased should be given its own record in the database, and each record should be given its own unique identifier, usually a number. For example, a utility could use the manufacturer's serial number as each meter's computer record identifier or could identify each meter according to the utility's own meter numbering system if the utility has one.

Meter data should be entered into the system at the time the meter is purchased or first installed in a customer's service or whenever any testing or repair work is done on the meter. These data can be entered manually or scanned into the system by the person receiving or repairing/testing the meter. This information is key for the utility's asset management/tracking system.

Additional factors that need consideration extending beyond the meter to other equipment and resources include pressure regulating, water quality monitoring, water loss reporting, and service records that include the ongoing maintenance and updates for the meter testing equipment. Certification of the equipment includes all components of the gravimetric and/or volumetric testing system.

The following list provides the basic data any record system should contain and includes information that may be found on the system.

Basic Meter Testing Information

- Meter size
- Meter manufacturer
- Meter model
- Meter type
  - Test selection (per requirements of AWWA meter standards)
  - Construction material
- Meter serial number
- Register model/type
  - Register identification (ID)
- Date of purchase
- Account number (to link with CIS)
  - Address (US Postal Service format)
- Type of test
  - Standard
  - Dispute/witness test
- Flow rates tested
  - Minimum rate
  - Intermediate rate
  - Maximum rate
  - Customized rates
- Results
- Tested by
- Historical
  - Testing history
  - Testing date(s)
- Water audit detail
- Service delivery point (an identifier that is independent of the meter/radio ID)
- Comments

Additional Meter Testing Information

- Water temperature at time of test
- Current location
    - Geographic information system (latitude and longitude)
    - Previous locations
    - Reason for relocation
- Repair history
- Repair data
    - Type of repair(s)
    - Materials/parts used
    - Repaired by
    - Labor time
    - Comments
- Billing information
    - Billing cycle
    - Billing route
- Test mode
    - On/off during testing

Additional information

- Water quality detail
- Water source (meter location [e.g., well, reservoir], not if recirculated or city supply)
- Type of service (automatic meter reading and advanced metering infrastructure)
    - Radio ID (in addition to the meter ID)
- Service type (may vary by utility)
    - Residential
    - Commercial
    - Multifamily
    - Irrigation
    - Nonbilled utility or city service
- Meter vendor reading/head end system
- Pressure zone
- District metered area
- Tap number (#) or tap map link

Additional Considerations

- Sources of data that are important to the utility
- Connection needed between different utility data systems
- CIS
- Water plant/treatment plant supervisory control and data acquisition system data
- Testing lab data

Mfr. No. Co. No. Make Size

Date Purchased Cost Style

INSTALLATION RECORD

Installed					Reasons for Removal	Removed	
Date	Reading	Name	Address	Tap No.		Date	Reading

TEST AND REPAIR RECORD

	Rate of Test			Repairs		Cost of Repair			
Date	Min. Flow	Inter. Flow	Max. Flow	% Accuracy Before Repair	% Accuracy After Repair	Mtl.	Labor	Tested By	REMARKS

Figure 8-1 Meter history record

- Internal intranet
- Nonmeter data such as
  - other considerations for the data system that extend beyond the meter and that are of value to the utility,
  - water usage reports for closed-loop operation, and
  - water quality reports for both "city supply" and closed-loop operation.

Sample reports follow the Summary at the end of this chapter.

PAPER METER RECORDS

Many utilities continue to maintain the data and meter details in a paper record system. Technology, economics, and data/operator integrity provide the pathway for an eventual migration to a computerized system, but the following documentation in the form of a meter history record (Figure 8-1) provides the necessary data that needs to be captured for a volumetric operation.

Meter test and repair documentation can be maintained with forms as outlined in Figure 8-2.

REPORT OF METER TEST AND REPAIRS

SHOP NO. ____________________

SHOP NO. ____________________

SIZE	MAKE	TYPE	NUMBER

METER READINGS	
AS RECEIVED	AFTER REPAIRS

REASON FOR TEST OR REPAIR					
STOPPED	FROZEN	LEAKING	TURN-OFF	COMPLAINT	PER. TEST
OTHER:					
DATE METER LAST RECONDITIONED					

CONDITION AS RECEIVED				CONDITION AFTER REPAIRS			
	QUANTITY				QUANTITY		
RATED G.P.M. C.F.M.	NET WEIGHT	GALS. CU. FT.	RESULT (%)	RATED G.P.M. C.F.M.	NET WEIGHT	GALS. CU. FT.	RESULT (%)
REGISTER GEAR	SPINDLE GEAR			REGISTER GEAR	SPINDLE GEAR		

REPAIR PARTS USED

REPAIRED BY	DATE	TESTED BY	DATE

REMARKS

(FRONT)

Meter No. ____________________
Meter Size ____________________
Meter Make ____________________
Date of Invoice ____________________

RECORD			
Set			
Date	Service No.	Removed Date	Tested Date

(BACK)

Figure 8-2 Report of meter test and repairs

SUMMARY

A best practice is to associate the data with a service delivery point. If the meter changes, your data do not need a complex retrieval process for analyzing nonrevenue water, diurnal and seasonal usage, and sites with multiple meters, sensors, transmitters, or other interrelated components.

A computerized meter test facility can provide a wealth of information when properly maintained. Although various systems are used for maintaining meter records, select systems should provide all the data the utility requires to monitor its operations. The system should provide detail suitable for data analytical evaluation (Figures 8-3 through 8-7). This provides the utility with the means to make informed decisions supported by credible and reliable detail.

Automated operation provides multiple benefits to the overall operation of the modern utility.

- It substantially reduces human error caused by replication of handwritten information. A review of almost any system of meter records will reveal duplications or more elaborate and detailed records than are needed. Addressing this problem will result in the elimination of unnecessary clerical work with only slight changes in the system or method.
- It reduces billing corrections, which minimizes the need to issue credits that reduce revenue.
- It provides valid, credible detail for water audits and water loss reports. See Figure 8-3 for an example of a report that can be provided to the end-user or the court.
- It provides valid, credible detail for billing dispute legal cases that stand up in court.
- It provides a foundation for future and/or customized meter testing data.

SAMPLE TEST REPORTS

Test Details

Test Date:	7/24/2025 7:58 AM	**Test Name:**	AWWA C700/C710 – Gal New – PosDis – ⅝″ x ¾″		
Job Number:	#WO1890	**Remarks:**			
Bench:	Bench1	**Bench Operator:**	Operator #1	**Units:**	Gallons

Subtest Details

Name	Tank	Line	Exp Rate	Act Rate	Exp Vol	Act Vol*	Low Tol	High Tol
Low Flow	10 Gal	½ in.	0.25	0.266	10.00	10.015	5.00%	1.50%
Mid Flow	10 Gal	1 in.	2.00	2.231	10.00	9.964	1.50%	1.50%
High Flow	100 Gal	1 in.	15.00	15.321	100.00	100.35	1.50%	1.50%

Meter Details

Meter	Subtest	Comp*	Start	End	Volume*	Accuracy	Pass
#1 – 93475582 Failed Meter "Y"	Low Flow	No	2029602.60	2029608.20	5.60	55.9161	No
	Mid Flow	No	2029608.20	2029618.10	9.90	99.3576	Yes
	High Flow	No	2029618.10	2029717.10	99.00	98.6547	Yes

In the Meter Details section, subtests shaded gray indicate the meter's accuracy was outside the specified tolerance for that subtest.
* Act Vol = The volume of water run through the meters during the subtest.
Volume = The volume of water registered by the meter during the subtest.
Comp = "Yes" if the meter has two registers, "No" otherwise.

Figure 8-3 Standard results test report (ANSI/AWWA C700/C710 Example)

Test Details

Test Date:	7/6/2025 9:58 AM	Test Name:	AWWA C715 – Gal New – Ultrasonic – ⅝″ x ¾″		
Job Number:	#WO1450	Remarks:	Gravity Feed Supply for Ultra-Low Flow		
Bench:	Bench1	Bench Operator:	Operator #2	Units:	Gallons

Subtest Details

Name	Tank	Line	Exp Rate	Act Rate	Exp Vol	Act Vol*	Low Tol	High Tol
High Flow	100 Gal	1 in.	15.00	15.164	100.00	100.334	1.50%	1.50%
Mid Flow	10 Gal	½ in.	0.40	0.435	10.00	10.01	1.50%	1.50%
Low Flow	ULF Tank	ULF	0.13	0.143	10.00	10.005	5.00%	5.00%

Meter Details

Meter	Subtest	Comp*	Start	End	Volume*	Accuracy	Pass
#3 – 21613908 Passed Meter "W"	High Flow	No	100404.784	100506.18	101.396	101.0584	Yes
	Mid Flow	No	100506.18	100516.202	10.022	100.1198	Yes
	Low Flow	No	100516.202	100526.577	10.375	103.6981	Yes

In the Meter Details section, subtests shaded gray indicate the meter's accuracy was outside the specified tolerance for that subtest.
* Act Vol = The volume of water run through the meters during the subtest.
Volume = The volume of water registered by the meter during the subtest.
Comp = "Yes" if the meter has two registers, "No" otherwise.

Figure 8-4 Standard results test report (ANSI/AWWA C715 Example)

Test Details

Test Date:	7/10/2025 9:53 AM	Test Name:	AWWA C700/C710 – Gal New – PosDis – ⅝″ x ¾″		
Job Number:	#WO1821	Remarks:			
Bench:	Bench1	Bench Operator:	Operator #1	Units:	Gallons

Subtest Details

Name	Tank	Line	Exp Rate	Act Rate	Exp Vol	Act Vol*	Low Tol	High Tol
Low Flow	10 Gal	½ in.	0.25	0.253	10.00	10.015	5.00%	1.00%
Mid Flow	10 Gal	1 in.	2.00	2.258	10.00	9.581	1.50%	1.50%
High Flow	100 Gal	1 in.	15.00	15.443	100.00	100.382	1.50%	1.50%

Meter Details

Meter	Subtest	Comp*	Start	End	Volume*	Accuracy
#1 – 14469282 Passed Meter "X"	Low Flow	No	161442.825	161452.813	9.988	99.7304
	Mid Flow	No	161452.813	161462.419	9.606	100.2609
	High Flow	No	161462.419	161561.779	99.36	98.9818
Weighted Meter Accuracy:			99.99			

In the Meter Details section, subtests shaded gray indicate the meter's accuracy was outside the specified tolerance for that subtest.
* Act Vol = The volume of water run through the meters during the subtest.
Volume = The volume of water registered by the meter during the subtest.
Comp = "Yes" if the meter has two registers, "No" otherwise.

Figure 8-5 Weighted average test report

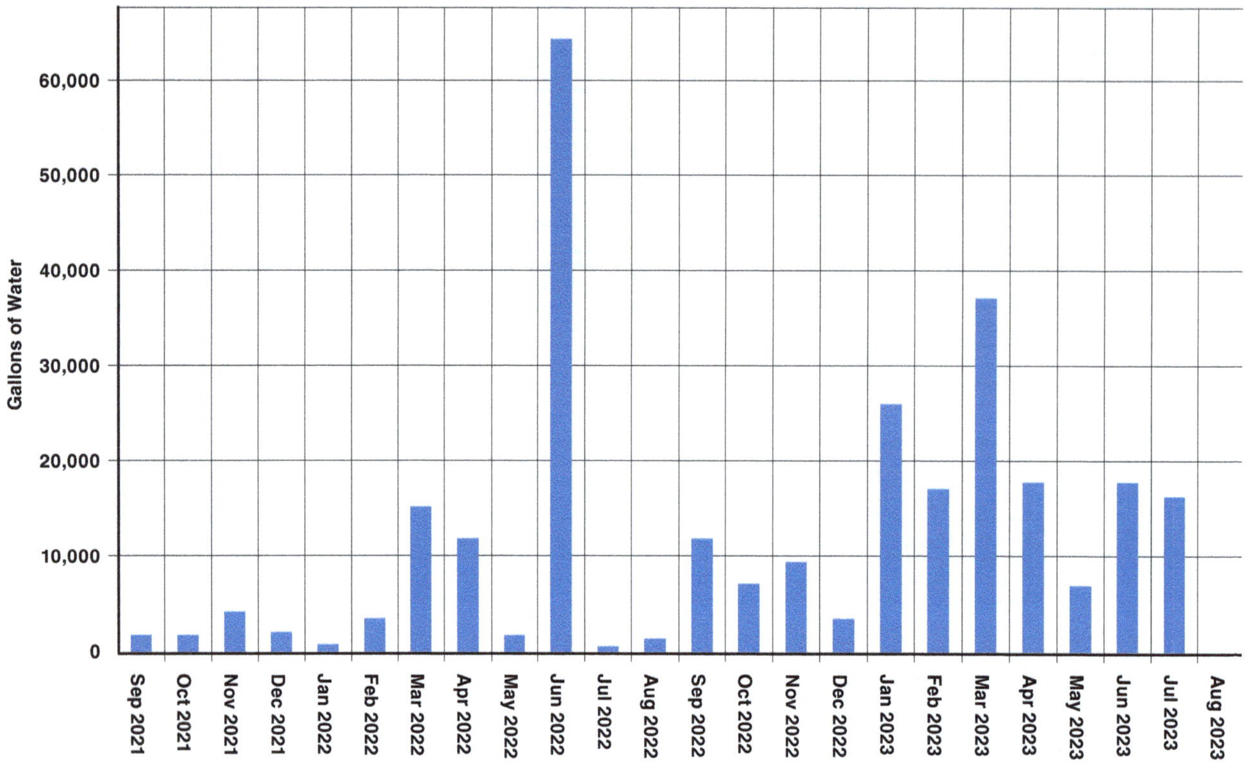

Figure 8-6 Meter test water usage report

Test Information

Test Date:	7/24/2025 8:16 AM	**Test Name:**	Sample Test	**Test Type:**	Meter Test – On Site

Water Quality Test Details

Alkalinity x 100 PPM	3.0000
pH	7.0000
Hardness x 10 PPM	8.0000
Iron x 1 PPM	9.0000
Copper x 1 PPM	2.0000
Lead x 10 PPM	4.0000
Fluoride x 1 PPM	0.0000
Mercury x 0.01 PPM	2.0000
Nitrate x 10 PPM	8.0000
Nitrite x 10 PPM	4.0000
Chromium x 1 PPM	6.0000
Bromine x 1 PPM	4.0000
Residual Chlorine x 1 PPM	9.0000
Sulfate x 100 PPM	8.0000
Notes	Sample Test

Figure 8-7 Water quality report

Chapter 9

Remote Registration

HISTORY

Water meter remote registration refers to the ability to read and/or manage a utility meter from a location that is not physically connected to the meter itself. This is typically achieved through various communication technologies, such as radio frequency, cellular networks, or internet connections.

Most of the water meters in North America are installed either in outdoor meter pits and vaults at some point along the service line or within a structure with limited access for the meter reader, such as in basements, mechanical rooms, or utility closets. Water utilities have been using various remote-reading systems to improve meter-reading accuracy and efficiency without having to gain physical access within structures.

During the last half century, water utilities have seen meter reading evolve from manual entry readings in a route book to automatic meter reading (AMR) systems to highly sophisticated advanced metering infrastructure (AMI) systems enabling the utility to add monitoring and control, optimizing operations and improving the services they provide. Recent developments in AMI systems allow the network to support other devices that are quite useful to the utility, such as pressure sensors, temperature monitors, and water quality monitors. Several important milestones marked this transformation from the basic manual system to the present-day, fully automated systems.

The first was the development of a remote-reading system that eliminated both lock-out problems and potential liability issues when reading meters located indoors. The first widely used remote-reading system was introduced to the water industry in 1957. It consisted of a self-generating pulse register and a remotely located electromechanical register. After a specific volume of water was registered, a pulse was generated and sent to the remote register. Remote-reading systems utilizing encoder registers with plug-in remote reading followed, along with various hybrid remotes for water meters.

The next step in the evolutionary process was the introduction of handheld computers (data collectors) to allow meter readers to enter readings directly into a computer memory. Readings are transferred directly into the utility's computer billing system, thereby improving productivity, accuracy, information gathering, and customer service, while eliminating manual data-entry and shortening the read-to-bill cycle.

The next development was onsite automatic entry of the encoded meter reading into a handheld device (by probe, wand, or other device). By adding an adapter to the handheld computer, the meter reading and account number are obtained by inserting the adapter into a remotely located socket or by touching a pad connected to an encoder register on the water meter.

The early 1980s brought a revolution to meter reading efficiency through mobile AMR systems, which are still a viable technology today for many utilities. AMR systems are typically operated in walk-by or drive-by modes depending on the collector(s) used by the utility. The encoder register on a mechanical water meter or the electronic output from a static water meter is connected to a radio frequency transmitter that broadcasts the meter readings and possibly other data—such as leak alerts, datalog, and tampering alarms—either on command or at programmed intervals. The datalog is typically hourly or daily meter readings stored for the preceding months. In walk-by mode, a portable, battery-powered collector captures the data from the endpoints and then stores the information for automatic downloading into the utility billing system software. In drive-by mode, a vehicle-mounted collector captures, stores, and downloads the data. Utilities may operate in a combination of walk-by and drive-by modes.

The 1990s saw the next great advancement in meter reading technology with the development of AMI systems that use permanently installed collectors, routers, or an existing cellular network to capture hourly or more frequent reading data along with meter alerts and alarms on a daily or more frequent basis. Those data are then transmitted to a server that is installed at either the water utility or is hosted by a cloud-based services provider. Typical backhaul methods for these data include Ethernet, Wi-Fi, or cellular modems.

Most utilities with AMI systems also use a meter data analytics software program to minimize both real and apparent water loss, to protect billable revenue, and to support hydraulic modeling. The utility may choose to provide limited access to the interval usage data through a customer web portal that assists the account owners in monitoring their water usage. A customer web portal may be used to inform the account owner of leaks at their property or other issues with their water service.

While information on selecting, installing, and maintaining an AMR or AMI system is outside of the scope of this manual, the utility is advised to closely follow the endpoint installation instructions from the AMR/AMI system provider in order to obtain optimum system performance. The utility is encouraged to actively maintain their investment in either an AMR or an AMI system network. For example, if an endpoint in an AMR system has not been read in two consecutive billing cycles or if an endpoint in an AMI system has not been read in 72 h, a technician should be dispatched to troubleshoot the problem.

Battery life is a major operating component of an AMR/AMI system. Utilities are encouraged to consider the coordination of battery life with meter life (accuracy) to minimize onsite customer visits. However, an AMR/AMI system is a depreciable asset over the term of its useful life. If, for example, an AMR/AMI component is replaced in year 15 of a 20-year estimated life, 25 percent of its remaining depreciation value is lost. Though a meter is also a depreciable asset for accounting purposes, the loss of revenue due to meter inaccuracy can far exceed its remaining depreciation value. The utility might find it is more cost-effective to replace an AMR/AMI system prematurely to minimize service call costs than to defer meter maintenance activities until the AMR/AMI system runs to the end of its estimated life.

CURRENT STATUS

As of 2022, two-thirds of all the residential water meters sold in the United States were equipped for some form of AMR or AMI, including a high percentage of commercial and industrial meters. Whereas early systems used cable or telephone networks, these systems have been surpassed by the ever-increasing use of radio frequency (RF) technologies. These RF systems are employed via walk-by, drive-by mobile, and fixed network, which is most applicable for large systems. RF has now become the preferred "information highway" of AMR systems, and it is this area of the meter-reading process that is the most rapidly advancing technology in the water-utility industry. Some AMI systems on the market today for water utilities support a smart-city concept, which is often defined as a municipality that uses information and communication technologies to increase operational efficiency, share information with the public, and improve both the quality of government services and citizen welfare.

BENEFITS OF REMOTE REGISTRATION

RF networks have enjoyed rapid growth due to their capability to provide benefits to the utility far beyond simple meter reading, whether they are deployed on all or key parts of the utility's distribution system. Some of the benefits of the new technology include the following.

- Increased sampling rate of data to enable
  - load profiling,
  - tamper detection,
  - leak detection,
  - no-flow detection,
  - backflow/reverse flow warnings, and
  - data needed for high bill investigation.
- Improved customer service:
  - ease of special reads,
  - elimination of estimated reads, and
  - hourly to daily data feeds enable customer water usage dashboards.
- Reduction of utility personnel risk during manual reading, connection, or disconnection to
  - reduce face-to-face interactions for shutoffs and
  - reduce field visits to troubleshoot high usage scenarios.
- Increased utility efficiency:
  - utility personnel concentrate on more critical tasks and
  - increased water system visibility, including
    - water distribution system metrics like
      - water volume and
      - pressure and
    - water quality metrics like
      - temperature,
      - chlorine,

- conductivity,
- turbidity, and
- fluoride.

The availability of these features varies between systems and depends on the design details of the systems and their components. For this reason, it is impossible to provide more specific details in this manual. It is recommended that utilities and other interested parties contact the various manufacturers for details on the features their systems provide.

Index

Note: *f* indicates figure; *t* indicates table

O

R

S

T

W

AWWA Manuals

M1, *Principles of Water Rates, Fees, and Charges*, #30001
M2, *Instrumentation and Control*, #30002
M3, *Safety Management for Water Utilities*, #30003
M4, *Water Fluoridation Principles and Practices*, #30004
M5, *Water Utility Management*, #30005
M6, *Water Meters—Selection, Installation, Testing, and Maintenance*, #30006
M7, *Problem Organisms in Water: Identification and Treatment*, #30007
M9, *Concrete Pressure Pipe*, #30009
M11, *Steel Pipe—A Guide for Design and Installation*, #30011
M12, *Simplified Procedures for Water Examination*, #30012
M14, *Backflow Prevention and Cross-Connection Control: Recommended Practices*, #30014
M17, *Fire Hydrants: Installation, Field Testing, and Maintenance*, #30017
M19, *Emergency Planning for Water and Wastewater Utilities*, #30019
M20, *Water Chlorination/Chloramination Practices and Principles*, #30020
M21, *Groundwater*, #30021
M22, *Sizing Water Service Lines and Meters*, #30022
M23, *PVC Pipe—Design and Installation*, #30023
M24, *Planning for the Distribution of Reclaimed Water*, #30024
M25, *Flexible-Membrane Covers and Linings for Potable-Water Reservoirs*, #30025
M27, *External Corrosion Control for Infrastructure Sustainability*, #30027
M28, *Rehabilitation of Water Mains*, #30028
M29, *Water Utility Capital Financing*, #30029
M30, *Precoat Filtration*, #30030
M31, *Distribution System Requirements for Fire Protection*, #30031
M32, *Computer Modeling of Water Distribution Systems*, #30032
M33, *Flowmeters in Water Supply*, #30033
M36, *Water Audits and Loss Control Programs*, #30036
M37, *Operational Control of Coagulation and Filtration Processes*, #30037
M38, *Electrodialysis and Electrodialysis Reversal*, #30038
M41, *Ductile-Iron Pipe and Fittings*, #30041
M42, *Steel Water-Storage Tanks*, #30042
M44, *Distribution Valves: Selection, Installation, Field Testing, and Maintenance*, #30044
M41, *Fiberglass Pipe Design*, #30045
M46, *Reverse Osmosis and Nanofiltration*, #30046
M47, *Capital Project Delivery*, #30047
M48, *Waterborne Pathogens*, #30048
M49, *Quarter-Turn Valves: Head Loss, Torque, and Cavitation Analysis*, #30049
M50, *Water Resources Planning*, #30050
M51, *Air Valves: Air-Release, Air/Vacuum and Combination*, #30051
M52, *Water Conservation Programs—A Planning Manual*, #30052
M53, *Microfiltration and Ultrafiltration Membranes for Drinking Water*, #30053
M54, *Developing Rates for Small Systems*, #30054
M55, *PE Pipe—Design and Installation*, #30055
M56, *Nitrification Prevention and Control in Drinking Water*, #30056
M57, *Algae: Source to Treatment*, #30057
M58, *Internal Corrosion Control in Water Distribution Systems*, #30058
M60, *Drought Preparedness and Response*, #30060
M61, *Desalination of Seawater*, #30061
M62, *Membrane Applications for Water Reuse*, #30062
M63, *Aquifer Storage and Recovery*, #30063
M65, *On-Site Generation of Hypochlorite*, #30065

M66, *Cylinder and Vane Actuators and Controls—Design and Installation,* #30066

M68, *Water Quality in Distribution Systems,* #30068

M69, *Inland Desalination and Concentrate Management,* #30069

M77, *Condition Assessment of Water Mains,* #30077

M80, *Biological Drinking Water Treatment* #30080

www.ingramcontent.com/pod-product-compliance
Ingram Content Group UK Ltd.
Pitfield, Milton Keynes, MK11 3LW, UK
UKHW052012041026
15590UKWH00008B/2302

9 781647 172985